T0184710

GEOGRAPHY, ART, RESEARCH

This book explores the intersection of geographical knowledge and artistic research in terms of both creative methods and practice-based research. In doing so it brings together geography's 'creative turn' with the art world's 'research turn.'

Based on a decade and a half of ethnographic stories of working at the intersection of creative arts practices and geographical research, this book offers a much-needed critical account of these forms of knowledge production. Adopting a geohumanities approach to investigating how these forms of knowledge are produced, consumed, and circulated, it queries what imaginaries and practices of the key sites of knowledge making (including the field, the artist's studio, the PhD thesis, and the exhibition) emerge and how these might challenge existing understandings of these locations. Inspired by the geographies of science and knowledge, art history and theory, and accounts of working within and beyond disciplines, this book seeks to understand the geographies of research at the intersection of geography and creative arts practices, how these geographies challenge existing understandings of these disciplines and practices, and what they might contribute to our wider discussions of working beyond disciplines, including through artistic research.

This book offers a timely contribution to the emerging fields of artistic research and geohumanities, and will appeal to undergraduate and postgraduate students and researchers.

Harriet Hawkins co-founded the Centre for the GeoHumanities at Royal Holloway, University of London, where she is a Professor of GeoHumanities. Her research focuses on the intersections of geography, art, creativity, aesthetics, and the imagination, and often involves collaborations with artists and arts organisations.

Routledge Research in Culture, Space and Identity

Series editor: Dr. Jon Anderson, School of Planning and Geography, Cardiff University, UK

The *Routledge Research in Culture, Space and Identity Series* offers a forum for original and innovative research within cultural geography and connected fields. Titles within the series are empirically and theoretically informed and explore a range of dynamic and captivating topics. This series provides a forum for cutting edge research and new theoretical perspectives that reflect the wealth of research currently being undertaken. This series is aimed at upper-level undergraduates, research students and academics, appealing to geographers as well as the broader social sciences, arts and humanities.

For more information about this series, please visit: https://www.routledge.com

GEOGRAPHY, ART, RESEARCH

Artistic Research in the GeoHumanities

Harriet Hawkins

Routledge
Taylor & Francis Group

LONDON AND NEW YORK

First published 2021
by Routledge
2 Park Square, Milton Park, Abingdon, Oxon OX14 4RN

and by Routledge
52 Vanderbilt Avenue, New York, NY 10017

Routledge is an imprint of the Taylor & Francis Group, an informa business

British Library Cataloguing-in-Publication Data
A catalogue record for this book is available from the British Library

Library of Congress Cataloging-in-Publication Data
A catalog record has been requested for this book

ISBN: 978-0-367-40615-8 (hbk)
ISBN: 978-0-367-55835-2 (pbk)
ISBN: 978-0-367-80000-0 (ebk)

Typeset in Bembo
by codeMantra

For those whose love and support is expressed by a skillful balance of faith and critical questioning

CONTENTS

ILLUSTRATIONS

Figures

Table

ACKNOWLEDGEMENTS

'It sounds a lot like doing art with your mates,' remarked a close friend on hearing about my research. In a sense he is not far wrong. This book is a product of many friendships, some of which existed before the research that grounds this book, and some of which emerged through it.

These ideas are the product of years of inspiring and generous discussions and practice alongside Flora Parrott and Marina Henfrey, Libby Straughan, Sallie Marston, Katherine Brickell, Dydia Delyser, Sasha Engelmann, Cecilie Sachs-Olsen, Rachael Squire, and Tania El Khoury, as well as Al Pinkerton, Alison Williams, Anne-Laure Amilhat-Szary, Annie Lovejoy, Anja Kanngieser, Aoife Kavanagh, Ben Murphy, Bergit Arends, Candice Boyd, Charlotte Veal, Chris Gibson, Christina Della Guistina, Clare Booker, Daniel Pavia, Danny McNally, Danielle Schreve, Deborah Dixon, Dydia Delyser, Giada Peterle, Eden Kinkaid, Ella Harris, Eva Barbarossa, Eric Magrane, Fearghus Ó Conchúir, Hattie Coppard, Helen Scalway, Iain Biggs, Ian Cook, Innes Keighren, Irene Hedigger, Jareh Das, Jen Bagelmann, John Wylie, Julian Ruddock, Karen Till, Lucy Mercer, Laura Price, Luce Choules, Matthew Flintham, Miriam Burke, Nadia Bartolini, Nelly Ben Hayoun, Paula Turner, Paulina Nordstrom, Pauline Ginard, Pete Adey, Perdita Phillips, Rachel Hughes, Rachel Pain, Rachel Woodward, Robert Hampson, Rosa Cerarols Ramiez, Rupert Griffiths, Ruth Catlow, Ruth Raynor, Sarah de Leeuw, Sue Ballard, Tariq Jazeel, Thomas Dekeyser, Tim Cresswell, Toni Luna, Vandana Desai, Varyl Thorndycroft, and Veronica della Dora, Wayne Chambliss.

This book is also a product of the last eight years I have spent in the Department of Geography, Royal Holloway University of London, where I arrived in 2012 having been told on many occasions, in a range of ways that I was 'too arty' and 'not geographical enough,' for various other geography departments. Not only did the RHUL department accept that my work was as much about the arts and humanities as it was about geography, they positively celebrated that and I was welcomed into a departmental research culture that encourages experimentation, and where generous and imaginative colleagues have spent decades working in innovative ways to bring together geography and creative practices. To find I no longer had to fight to have my perspectives and concerns accepted was a real gift, and without question created the space and context for the thinking and practicing that guides this book. If my early work on art and creativity would not even have happened without the guidance of Stephen Daniels and Nick Alfrey and colleagues at the University of Nottingham, it would not have continued within the environment created by colleagues and mentors at RHUL, especially within geography. Phil Crang, David Gilbert, and Felix Driver, thankyou, the selflessness with which you commit to creating a research culture in which others can flourish is quite remarkable.

Integral to this research has been the courageous work of the group of PhD students at RHUL: Ben Murphy, Bergit Arends, Bethan Lloyd-Worthington, Clare Booker, Danny McNally, Flora Parrott, Gloria Lowe, Ella Harris, Hattie Coppard, Jareh Das, Laura Price, Lucy Mercer, Miriam Burke, Nelly Ben Hayoun, Rupert Griffiths, and Tania El Khoury; your abilities to push geographical scholarship and practice in new directions are endlessly inspiring.

My heartfelt thanks go too, to all those who so generously shared their time and experiences with me over the course of this project, including through the formal interviews; without your willingness to share your experiences this book would be far less rich, if it existed at all.

The research that underpins this book has been funded by the generousity of a range of funders over the years. These include the Arts and Humanities Research Council (UK), The National Science Foundation (US), The Swiss National Science Foundation, the Royal Geographical Society (with IBG), and The Leverhulme Trust. The time spent writing the book was funded by a Phillip Leverhulme Prize, which allowed freedom to practice, write and think, in a way that is almost unbelievable in the context of today's funding climate. During this time two exciting and innovative early career geographer-practitioners covered my teaching at Royal Holloway, Cecilie Sachs-Olsen and Sofie Narbed, and inducted undergraduates and masters students into creative practice-based geographies in exciting and engaging ways that extended the craft of geography teaching, and helped ensure the sustainability of these ways of doing geography.

None of the written material has appeared anywhere else, and I thank the owners of the images (listed in their captions) for granting me permission to use them. The text has benefited from the support and patience of editors and staff at Routledge, including Faye Leerink, Nonita Saha and Jeanine Furino. Many of the ideas have been presented over the years in a range of conferences and seminar contexts, and have benefitted greatly from the generous and probing comments of scholars around the world and working in many contexts. Perhaps, most testing has been taking this work out of its UK-US context and offering it up to audiences researching in other contexts, including, at the Universities of Shenzhen and Guangzhou; discussing it with colleagues from the Universities of Konkuk and Santiago, as well as reflecting on the geographies of practicing geography with scholars working at Ecole Normale Supérieure, Paris; Pompeu Fabra University, Barcelona; ETH and ZHdK, Zurich and Tallinn University.

As ever, nothing would ever be made, and certainly never finished – books, art, or otherwise – without the support of my family, Anne, Louis, and Joseph Hawkins, and Pat Jarvis, practitioners of creative geographies in many forms.

INTRODUCTION

The elephant in the room...

'The elephant in the room- the arts' was the opening gambit of one interviewer on an interview panel I faced as I was coming to the final phases of writing this book. This was a panel of social scientists, and my proposal was to study subterranean spaces – key to our environmental futures. I wanted to tackle the challenges of sensing, imagining, and speculating on these spaces through an approach whose cornerstone was a combination of arts practice-based research and creative research methods of the sort increasingly found within social science. Over the next twenty minutes or so, I was called on to defend this artistic approach as science, to explain how it was that practice-based work and creative methods would produce research, and how I expected this to change both science and the arts. I was pushed to demonstrate how this was not just applied research or public engagement, defending these approaches as 'basic science,' as actual knowledge making. It was ultimately an invigorating experience. I did not expect to get the grant, but I did come away with a renewed sense of purpose around the place of art practice within social science research and the need to argue for it more clearly and persuasively. Working as a cultural geographer in a UK context, I had perhaps become too used to arts practices being inherently valued as research. I am lucky enough to work in a geography department with a long history of collaboration with creative practitioners, arts-based methods are a feature of many undergraduate and postgraduate curricula, and writers, artists and dancers often appear as PhD students, colleagues, and visitors. Indeed, in calls for papers for major disciplinary conferences 'alternative formats' has become almost a standard line, populating geography conferences with poetry readings, visual essays, film screenings, installations, and dance. Yet the panel's robust questioning reminded me that what I had come, perhaps somewhat lazily, to accept as valued and justified research practice was still for many the site of debate. It was not clearly science – nor research, as perhaps this panel meant it – nor knowledge making in its own right, nor even really part of a methods tool box; at best arts practices' relationships to research were seen as a form of dissemination and perhaps public engagement. It was hard to articulate in the three minutes I had been advised I had for each question, how the arts were a form of knowledge making, how creative practice could be research.

I begin with this anecdote because it frames some of the key questions from which this book emerges, namely how exactly has artistic practice come to be a form of geographical research? How does a belief that art is a valid form of geographical research enquiry, not just a cultural form for geographers to study, require that we revisit our understandings of the arts and of geographical research? How might an appreciation of the coming together of the arts and geographical research require us to shift understandings of both? This volume attempts to gain some purchase on these rather expansive questions. This anecdote, importantly, does not just frame key questions, it is also indicative of this volume's approach. Very few finished art-works – such as films, installations, or paintings – appear in this book. Rather, I seek ways into these somewhat daunting questions through an ethnographic approach that attempts to make creative practice and research relations felt from within the often everyday spaces, doings, thinkings, and sayings that constitute these ways of working.

This research on research is informed by my decade and a half of working at, and on, the intersection of geography and the arts. This has enabled the assembly of threads of thought and experience around a series of sites key to these forms of knowledge production: field, studio, lab, community, residency, page, thesis, and exhibition. It has also driven a recognition of shared concern with what might be best described as a stagnation of the field. On the one hand, are geographers for whom geography-arts relations require further critical attention, either because they have reached somewhat of a still point, stuck in a hiatus of criticality, or because their cult-like status has foreclosed more critical discussions.[1] On the other hand, is a seemingly parallel sense from within creative practice-based research – from fields as diverse as creative writing, drama, design, and visual art – that much needed critical discussion is often shut down. For many this is a function of the primacy of understandings of research, and the terms and conditions of its practice and judgement, that are set by the cultures of Higher Education that often foreground science, and are ill-suited to these creative ways of doing research.[2]

To elaborate. Cultural geographers have long taken the diverse creative practices of others (from painters, to poets, musicians, and film-makers) as their subject. Yet, while this offers vital context, here I am more concerned with the growth in the number of geographers 'doing' creative practices as part of their own research methods.[3] The influences and instigators of this so-called 'creative turn' (or re-turn as some argue acknowledging the histories of these practices) are varied. Even a cursory roll-call would take in; visual anthropology, participatory geographies, and various evolutions of humanistic perspectives.[4] It would include diverse forms of non-representational theories and their attentiveness to concerns of practice, embodiment, and affect, and their resolute commitment to an experimentalism profoundly informed by the arts and creativity. Equally varied, as elaborated further below, are the appreciations of creativity that intersect in these discussions. We find creative practices as a circumscribed body of skilled, 'high' art practice, as amateur and everyday practices, as well as those understandings that situate creativity as part of worldly becomings.[5] It would also, interestingly, take in work happening around the world at a range of scales. As well as those individuals and collectives working across regions with strengths in cultural geography, including Europe, USA, Canada, and Australia, we also see the emergence of large (often well-funded) research centres across Asia that situate creative practices as part of their address of key geographical themes, from mobility to Earth Futures.[6] Importantly, Geography as a discipline is not alone in this creative turn, indeed, we see similar work emerging in anthropology, archaeology, sociology, politics, and international relations as well as history.[7]

If geography's creative turn is rich and varied, the 'research' turn within art offers an equally diverse expression of the growing force of research within the world of creative and artistic practice.[8] Without question, 'research' of a form has always been integral to artistic practice. Yet in recent years, these relations have been given new direction and force by the evolution of creative practices and the political economies of the art world, not least, its shifting relationship with Higher Education around the globe.[9] This has seen arts education and artists folded into wider university systems, including within the growing force of practices of interdisciplinary and concerns with public engagement. Yet artistic research should not be solely understood in terms of the relocation of artists within the academy. For there has also been an emergence of what we might call a research aesthetics. By this I mean a varied set of practices that includes, the growth in the use of research methods – from ethnography to archival work – as artistic mediums, the increasing use of notes, brainstorming documents and readings as part of installations or exhibitions (a kind of installation of literature reviews or a visualisation of the conceptualisation of the works), as well as the presentation of volumes of information as exhibitions with the effect of casting the audience as co-researcher.[10] If some have found these myriad intersections of research and practice to be productive – generating new ideas as well as points of critique (for the academy as much as for the arts) – others have been thoroughly resistant.[11] There has been concern about instrumentalisation of the arts, and the reductive understandings of artistic research that evolve under the force of the evaluative logics of global Higher Education's audit cultures.[12]

Geography, Art, Research emerges from the midst of these diverse and intensifying energies of the relationship between creative practices and geographical research. My intention is not to collapse the differences within or across the creative turn and the research turn. Indeed, as the discussion across the chapters will clearly demonstrate, the practices, philosophies, and ethics of these two turns are as internally diverse as are the differences between them. Nor am I looking to offer some kind of hierarchy or typography of these practices – valuing say professional over amateur, collaborative over solo, avant-garde over conservative, knowledge making over public engagement (nor of course are these binaries). Instead, from amidst their specificities I want to build a critical account of these ways of working that responds to the shared conviction of scholar-practitioners working in these ways that, in the midst of a period of exciting growth, we risk a critical hiatus. Looking to the wider field of creative practice research relations beyond geography, there is dismay that the popularity of these ways of working has not inspired greater critical thinking and theorising.[13] If some are vexed by the enduring traction of out-dated understandings of research (based on a model more suited to natural sciences or an objective social science), others are exercised by too of-the-moment faddishness for interdisciplinarity, impact, and certain research themes.[14] The result, whether too dated or too fashionable, is to risk compromised creative practices and compromised research. For others, the fetish for 'being creative' (shared of course with neo-liberal corporate managers and city planners) has resulted in power-relations, and a sense of respect and care for difference being overlooked. There are concerns that, driven by the rhetorics and demands of funders, institutions and the 'star culture' of academia, creative practice and research relations are really just performative. For all their possibilities, in the end, the fundamentals of our research cultures, politics, and practices are unchanged.[15] For others, university-based academic cultures are fundamentally ill-equipped to support these ways of working. There is a lack of infrastructure (whether buildings and equipment, or social networks for support and critique), and institutional cultures and contexts (from funding mechanisms to the formats of conferences and modes of evaluation) are too rigid.[16] Research in these cultures is seen as plagued by neoliberal creativities and dogged by textual imperialisms and latent scientism that demands written outputs, and sees in the arts application and engagement rather than 'basic' research.[17] Yet, despite these challenges, there is often a note of hope in the midst of many accounts. Indeed, for many practitioners I spoke to who have found a home within geography as artists-in-residence, as doctoral students, as post-doctoral researchers, and in some cases as academic staff, the discipline was not without these problems, but seemed to offer a positive 'home' for practice, and important, is a research culture largely open to change.

It should perhaps be of no surprise that the coming together of creative practice and geographical research presents some significant critical challenges. After all, centuries of art theory and history have been built on the back of the demands that evolving artistic mediums– from sculpture, to installation, to participatory art – pose to accepted understandings of how art is made, how it is consumed by its audiences, and the terms upon which its critics understand and judge it.[18] Indeed, the research turn within art might be understood as a continuance of much of nineteenth- and twentieth-century art's imperative to challenge what we understand art to be and thus how we can critique it. In other words, is the accuracy of representation the primary achievement of a piece of art? Is the harnessing of aesthetic force – of beauty or the sublime – a core critical concern, or, should we be judging art according to the terms of the avant-garde (is it transforming how we understand art)? In short, the coming together of creative practice and research (if we can even think of it in this way) requires that we continue to query the forms of critique that are appropriate, that we continue to hold together multiple appreciations of what it is that art might do and be. Should we be seeking from the coming together of creative practice and research, for example, the highest forms of aesthetic quality and production values, or should we be seeking avant-garde art, what is the latter in this context? What about good research? What is good research in this context? Is it acceptable to trade one off against the other? Great art but only acceptable research?

These are not just questions about art, but also questions that speak to how we understand research. After all, our changing understandings of what constitutes geographical knowledge has long been closely tied to the methods we use to make it, and the forms through which we consume it. We might think, for example, of Mayhew's entwining of the shifting forms of geographical writing and print cultures with the movement from a "Capes and Bays" Geography to a more scientific discipline, manifest on the page in the move away from encyclopaedic and gazetteer forms.[19] We could think too of how the rise of emotional and affective geographies with humanist, feminist, and non-representational theories, shifted the ground across which we understood questions of empirics, of forms of output, and so on, drawing to the fore questions of felt or practiced knowledge.[20] As such, it should come as no surprise that from within both creative practice-based disciplines and within geography there is a need to evolve a new critical framework for thinking about the relations of creative practice and research.

Geography, Art, Research's response to much needed criticality for this field is to experiment with a geohumanities approach. GeoHumanities names another recent intensification of working across, amongst, within and, for some, beyond disciplines in the making of geographical knowledge.[21] Folding relations between geographical research and creative practice within it, as a term geohumanities recognises the value of arts and humanities perspectives and practices for geographical knowledge making.[22] Of course, these have long been integral to geography as a discipline, but can tend to get written out in the context of the histories of the discipline as a science, even if this is a science understood to always include other ways of knowing.[23] What is more geohumanities – as it is enacted in this text – offers the means not just to gather texts and concepts from across disciplines, but to enrol practice and methods too as a means to respond to the questions posed by the diverse comings together of geographical research and creative practices. It seemed only fitting to meet a set of practices that so resolutely refuse to be contained within disciplinary boundaries with a thoroughly rich admixture of inspirations and contextual framings, although this brings, as I will elaborate on, its own challenges.

Experimenting with the geohumanities this book evolves three, and inter-woven, trajectories. The first explores some of the key sites of the coming to-gether of geographical research and creative practice. It does this by way of eight site-based chapters exploring, in turn, field, studio, laboratory, community, res-idency, thesis, page, and exhibition. Myriad historians of science and scholars of the history and philosophy of geography have been clear: we need to put 'science in its place.'[24] If we need to appreciate how scientific knowledge and the geographies of its production, consumption, and circulation are fundamentally co-constitutive, so too must we appreciate the value of putting creative practice and geographical research relations in their place. Indeed, cultural geographers have long appreciated the geographies of art, film, literature, and so on, entwine the geographies within the frame, page, and screen, with those of the work's pro-duction, consumption, and circulation (the studio, the site, the exhibition, the catalogue).[25] Thus, the chapters will interrogate the value of 'placing' knowledge made at the intersection of geographical research and creative practices. The promise of thinking geographically about creative practice-based research, is like thinking geographically about science, with the latter being 'too rich, too fertile, too intricate' to be reduced to a classificatory system that could offer anything more than a partial and tentative mapping of the geographies.[26] As such, this is less the mapping of a field's geographies than it is an attempt to appreciate and as-sert the critical value of attending to the geographies of the making of knowledge at the intersection of geographical research and creative practices.

The second trajectory traces the understandings of geographical research, and of arts and creativity that emerge from the midst of the eight sites. There is a well-recognised danger, as Felix Guattari articulated in a co-written report for UNESCO in 1992, that interdisciplinarity is simply an 'abracadabra word.'[27] In other words, a linguistic flourish, a flashy performative gesture that changes little in practice. I have already elaborated on the concerns the art world and geography share that we have got somewhat carried away by these practices. Indeed, have we just created a new category, grouping together methods that enable us to tick boxes, claim something new and 'reassure one's institution and oneself' that one is clearly operating 'at the cutting edge?'[28] Increasingly com-mon concerns are expressed that we have failed to address how 'the process of art and creative practice alter[s] what we might think of as research?'[29] I respond here with an empirical specificity that uses geographically situated small stories and the sometimes incidental details of daily practices to explore the understand-ings that emerge in the midst of these relationships.

The third trajectory considers how the sites at which geographical research and creative practices come together – be they those of the studio, the field, the thesis, or the exhibition – are themselves being (re)imagined through these couplings. The co-constitutive relationships between say the field and the meth-odological practices and epistemological foundations of geographical knowledge making, or the studio and the forms of art making and understandings of art, are well accepted. Yet, as has been made clear with respect to science, there is a need to move beyond merely producing 'spatial narratives that are constitutive of (and constituted by) the scientific facts they are about,' to push further at how these scientific practices also 'serve to question pre-existing understandings of space and the social relations sustained within them.'[30] We need, in short, to appreciate how the sites of science (here read geographical knowledge making) are being redefined and contested by these relations between creativity and research.

In following these trajectories, *Geography, Art, Research* both contributes to developing discussions about geographical research and creative practices, but is also of relevance to wider experiences of the research turn, and the creative turn within other disciplines. The remainder of this introduction has three priorities. First, to address the complex and unsettled ideas of creativity at work in this volume and how they play out in tensions within the creative and research turns. Second, I want to detail the experimental geohumanities perspective that I am exploring here, including the methodological approach that grounds the eight empirical chapters. Finally, I will offer some structural guidance and signposts for the remainder of this volume.

Creativity?

Managing this book's multiple threads was tricky for many reasons, but perhaps most consistently troubling was my going back and forth with the terms art, the arts, and creativity. Creativity within geography is, for some, as specific as it is for others capacious, as much valued as it is maligned – even my own department has managed to generate volumes titled both 'For' and 'Against' creativity.[31] As such, creativity is, as many others have argued, in need of some careful thought.

Creativity's multiple valences can make it stick in the throat. We live, Osborne suggests, in a 'veritable age of creativity.' Not least due to creativity's pervasive place in economic life in the form of the creative industries and creative economy, and the rise of the so-called 'creativity script' as a form of urban planning and governance shaping the look, functionality, and feel of cities around the world.[32] Within academia, the rise in popularity of creativity, less an aspiration than an imperative some claim, a brand rather than a concept, might be seen as Higher Education's own working through of neo-liberal creativities.[33] Yet in recent years many geographers, including diverse cultural geographers and non-representational theorists, have got beyond the gall of creativity to sit with its productive 'manyness.'[34] This takes in creative economy and industries, cultural geographies of art, literature and film, and so on, as well as to diffuse understandings of everyday or ordinary creativity, whether that be hobbies, home decoration, or gardening.[35] Further, there are also those diffuse, not-solely-human, senses of creativity, which speak of other worldly epistemologies that resist creativity as a distinct set of practices and instead see it as a generalised quality of worldly becoming.[36] Of course, there is also, as is my focus here, creativity as a methodological force and a research practice, enacted both in terms of distinct creative practices, e.g. film-making, as well as a general 'creative' approach to research.[37] To get anywhere with creativity then, is to be at peace with its many forms, to find productive difference and tension across its variations.

Having said that, the 'manyness' of creativity at play across this volume foregrounds certain forms. Primarily a focus on how diverse creative practices – from art, to dance, performance, creative writing, and so on – find their place within geographical research, whether as research methods used by geographers or as practice-based research approaches of artists, dancers, writers, and so on. Most of the case studies that populate the eight chapters foreground artistic practices, if such a medium-based distinction still makes sense. Driven by my own training, I mobilise art history and theory perspectives, rather than the perspectives of either another medium – such as creative writing – or of a single philosopher. Important to note however, is that art today is often understood as a 'post-medium' practice.[38] In other words, the figure of the artist as a practitioner of a single medium – a painter or sculptor, say – has largely been replaced by the artist as a creative practitioner in possession of diffuse practices, and whose projects may well draw on those of other artists and craftspeople with whom they may collaborate, or might just simply commission to fabricate certain aspects of their work. Such post-medium practice intersects with other creativities across the text, including those of geographers learning new creative skills and intersections between Higher Education and the wider creative and knowledge economies. I don't pretend these slippages in practice, in forms of creativity, and also in my choice of terms are easy, but trying to deny them, or sort them too clearly, seemed, in the writing of them, to cut against the complexity of the field I was working through. This is an issue the conclusion returns to, to try to seize in more detail.

Two turns

In order to gain some purchase on the rich variations of creativity that guide this text, I want to focus in on two turns: the so-called 'creative re-turn' within geography and the 'research turn' within creative practice. To talk of intellectual turns is, as I have explored elsewhere, a controversial thing to do.[39] It is to negotiate the politics of knowledge production, its fashions, fads, and jostling for name recognition as well as to grapple with imaginations of how we know. These might be imaginations of paradigmatic tipping points, or of the slower, more hesitate progress of temporary diversions from paths of query, or of the ever-running course of currents of knowledge production that disappear below the surface, out of sight, only to reappear again somewhere else at some other time.[40] Beginning from turns is risky, and I am worried about getting bogged down in overly descriptive accounts of vast fields, or debates regarding scale, scope, and form. I am also concerned to negotiate any sense of two distinct, ready-made fields, making something akin to a coherent 'turn' towards each other. The picture is, as the chapters unfold, of course, far messier. Yet turn talk does summon a certain sense of the force and purchase of these trends, their energies, and their gathering pace, it also helps usefully to tether some of the many ways creativity is at work across this volume.

Geography's creative turn

If creative practices have been long been an integral, if perhaps oftentimes over-looked, part of geographical scholarship, the idea of the 'creative (re)turn' within geography expresses that sense of the intensification, over the first few decades of the twenty-first century, of the place and value of creative practices within geographical research methods.[41] Many are keen to ensure that the histories and trajectories of creative practices within geographical knowledge making don't get lost amidst the excitement over the current value and growth of these prac-tices.[42] It is important, for example, to reflect on the place of artists on board ship in the Age of Exploration, or to revisit the place of creative practice within humanistic geographies, including Donald Menig's rallying cry for geographers to follow creativity where-ever it may lead.[43] There are by now, myriad accounts of geographical projects that deploy creative practices as research methods. These variously hail from participatory geographies, visual and sensory anthropology, or non-representational theories, and deploy a rich range of practices – filming, drawing, dance or music – in the service of an equally diverse range of substan-tive topics, from the environment, to migration and austerity. There is a small but increasingly numerous suite of edited collections, monographs, and PhD theses that explore these creative methods and approaches.[44] Across these various practices and perspectives three tensions can be identified, and it is these I want to address.

The first tension emerges from the appreciation of geographical research methods as forms of both reportage (of an existing world), and of world-making. This is best seen in the rich range of discussions about visual methods – from photography and video expanding into a broad and rich attention to prac-tices of drawing, cartooning, and scribing as well as increasingly technological work with drones, amongst other devices.[45] Such methods negotiate an episte-mological legacy wherein the image is accorded a status as truth-teller, and as information-rich – a picture is worth a thousand words. As such, image-making and collecting, whether by the researcher or with participants, is understood to create a better, fuller account of things that might be overlooked or missed in the moment, or of sensory experiences that might be hard to put into words.[46] By contrast, non-representational theories and so-called new empiricism asks us to consider creative practices as more than just a means to grapple with the affective, emotional, and sensory aspects of subjective experience.[47] Rather than offering access to the subjective aspects of life-worlds, creative practices are appreciated for opening us onto an epistemological stance based in the ongoing emergence of researcher and world.[48] This empirical unmooring replaces what we could think of as an extractive model – a going out into the world, collecting data and returning to write up/ report – with an experimentalism that uses creative practices to create conditions for intersecting research and world in ongoing and emergent ways – creative practices partaking then in world-making.[49] Such in-junctions towards creative experimentalism require a fundamental rethinking/ re-practicing of geographical research by way of the arts with, as the chapters elaborate potentially significant impacts for how it is we formulate ideas of empirics, or research methods, analysis, and writing up.[50]

A second interesting tension emerging within geography's creative turn concerns questions of skill. There are several dimensions here, do geographers develop these creative skills themselves, or do they collaborate with other more skilled practitioners? Does it matter how skilled you are at a creative practice? How, in turn, do we need to, or come to, appreciate the skilled practices of the geographer? Further, what does critique or, better, evaluation mean here, and what skills do we need, as a discipline, in order to conduct such evaluations? A healthy crop of discussions begin from auto-ethnographic accounts of geographers who lack skills, recounting research experiences that seem shaped by a certain discombobulation when faced with requests to dance, paint, draw, or develop taxidermy.[51] Much has been made of the value of learning to do and also of the need to embrace and value the amateur. Yet other geographers have many years of training and experience, and are highly skilled at those practices they choose to embrace within their research, or choose to probe the skills and practices of collaboration.[52] Linked to the rise in questions of craft, skill has also drawn attention for its ontological and epistemological dimensions, requiring that we think about as Merle Patchett and Joanna Mann describe, practice, process, technicity, politics, and ecologies.[53] Emerging from across these diverse discussions is the critical potential of skill, the need to respect skilled practice, but also to not assume that a lack of skill negates the value of practice. Indeed, we need to perhaps be more attuned to the different capacities of skilled and unskilled creative practice, as well as the value of collaboration when we think through relations between creative practice and geographic research.

A third tension increasingly central to geography's creative turn concerns politics – long a thorny issue when it comes to reflecting on cultural geographies.[54] For those scholars working in the tradition of participatory research, diverse creative practices, from those of video-making to theatre, story-telling, cartooning, and photography offer a means to extend the practices of participatory research.[55] Creative practices – whether developed by geographers alone or in collaboration with professional artists – offer participatory geographers a means to engage hard to reach communities, to raise difficult issues, and interestingly, to enable the social change that sits at the heart of participatory approaches. Yet, if some geographers' engagement with creative practices is thoroughly political, and indeed enables a negotiation of power relations, others have perhaps been less than aware of the power relations inherent in their creative practices. Indeed, as Sarah De Leeuw and colleagues observe, a 'tendency to "do good" or work with the "best intentions" can, if not critically reflected upon (re)produce and entrench norms of ubiquitous white-centric, heteronormative, patriarchal colonialism.'[56] As they assert we must remain vigilant around questions of, 'for whom work is being produced, who benefits, who (or what) is being left out, and how might we develop frameworks for critically examining creative arts- and humanities based ways of knowing and being.'[57] Further, we have a tendency as Sage Brice recounts, to overlook some of the more troubling, dark histories of our creative practices, and have, as work by Amanda Rogers and others make clear, often reproduced western-centric forms of creative practice and theoretical concerns in our approaches to these methods.[58] Overlooking the politics of these methods is also concerning for their ethical practice within the academic context. Whether this concerns a politics of citation, or relates to what Heather McLean, Divya Tolia-Kelly, and others have critiqued as the tendency for academic worlds to capture the labour of creative practitioners for their own ends – principally to offer forms of public engagement and impact, to enable interdisciplinarity and to offer edgy forms of cultural capital.[59]

These three tensions run throughout the empirical chapters which follow. In some cases they are elaborated upon, elsewhere they fall away. I am less seeking to resolve them, than I am to explore their force across the field, and how they interact with perspectives from across the art world, it is to these perspectives I will now turn.

The arts' research turn

If it needed confirmation, the place of 'research' within the contemporary art world was reinforced in 2012 when dOCUMENTA (13) – the five yearly 100-day art festival held in Kassel in Germany attended by a million visitors – was 'dedicated to artistic research.'[60] The art establishment's acceptance of research was further evidenced by the launch, in 2015, of a 'research pavilion' at Venice Biennale, the cornerstone of the global art-world exhibition circuit. Curated and funded by University of the Arts Finland the themed pavilion bucked the trend for Venice to be dominated by nationally organised displays, instead uniting international artist-researchers in an exhibition and a series of events designed to explore the myriad forms of artistic research.[61] The theme in 2019 was 'Research Ecologies,' bringing together reflections on transdisciplinary collaborations as social ecologies with conversations and artworks about the environment. As well as confirming the acceptance of research within contemporary art world discourse and practice, these activities also demonstrate how research within the art world is marked by a proliferation of definitions and practices, and a pervasive confusion about how, if at all, to police these ways of working. At best this ongoing negotiation of research is a productive focus for reflection; at worst it is the subject of fuzzy, obfuscated, and abstract debate.

To talk of a 'research turn' in art is for many somewhat odd. In the main because many artists consider research an ontological condition of art making. Pablo Picasso, for example, observed, 'paintings are but research and experiment. I never do a painting as a work of art. All of them are researchers. I search constantly and there is a logical sequence in all this research.'[62] Yet, just as there are diverse currents of 'creative turns' within geography and tensions across these, so there are too within the arts' research turn. I want to explore these through the lens of tension between creative practice and the 'Academy,' and what I will call research aesthetics.

William Hogarth's 1745 engraving 'The Battle of the Pictures' suggests that tensions between something called 'The Academy' and something called 'Art' is nothing new.[63] Hogarth presents the serried ranks of copies of the old master's (academic art) engaging in battle with the so-called 'avant-garde' (a series of canvases of Hogarth's own works, everyday street scenes, and interiors that are leaving his studio, escaping the easel and flying through the air). The avant-garde's flight from earth, order, and easel is a representation of breaking free from the regimentation of the Art Academy's regimes – a process of twinned aesthetic and social liberation. While there are, of course, fundamental differences between the art and the 'Academy' of Hogarth's era, and those of today (the breadth of art forms, the nature of the academy to name but two major ones) there are valuable echoes too. Not least of which is the vexed relationship of the Academy with cutting-edge art, and the relations between art and various forms of commercialisation and instrumentalisation, whether by the market, or more recently by Higher Education.[64]

A popular origin stories for art's research turn lies in the complicated relationship between arts education and the various manifestations of neoliberalism within universities around the world.[65] The global trend over the last decades for the relocation of art schools, music and dance academies, and so on into university contexts has brought about new sets of relationships between creative practices and university-based understandings of research.[66] The tensions within these relations often play out in the complex definitional debates that mark the field – whether it be practice-based or led research, artistic research or art as research, or research-creation – the understanding varies often by nation and institution as well as by individual.[67] These discussions often find their greatest friction at points of judgement. Whether that be the expectation that artists should have PhDs, driving the growth and evolution of the practice-based PhD (discussed in a geographical context in Chapter 6), or in the context of practices of promotion, audit, and funding. What many of these contexts have in common is a clash between what is understood to be research in an arts context and the understandings and practices of established and powerful (often scientifically or at least social-scientifically based) research cultures.[68]

So pervasive has this relationship between art and Higher Education become, that James Elkins, a leader in the field, suggests there is more literature on the practice-based PhD alone, than could be read in a life time.[69] A roll-call of topics would consider the wealth of debate about publication formats, about the nature of impact and public engagement, as well as wider questions of judgement and evaluation, of issues around the form and nature of arts Higher Education, and in some cases point-blank refusals that creative practices could constitute research. Indeed, the rise of the artist in Higher Education has become so pervasive that new sociologies and ecologies of art have been claimed, adopting the contemporary labour form of the hyphenated worker – academic-researcher-artist.[70] As such, reputations are built less through 'establishing a role and mastery' of a chosen medium, and are instead about a diffuse set of often immaterial labour practices. Here lecturing sits alongside, 'putting on shows, publishing (within and beyond academy), organising events, teaching, networking, maybe belonging to one of more collectives, even adapting more pseudonyms.'[71]

For some creative practitioners I spoke to, situating their practice within the university system offers a vital means to sustain it, providing a space in which to make blue-skies work, freed from the vagaries of the commercial art market and dealer system. For others, it is to subject practice to another set of market and regulatory forces, and in doing so to risk compromising the quality of the art, and the politics and ethics of the practitioner. Some see the need to give students numerical marks, and to rank and judge their colleagues on criteria emerging from managerial cultures more suited (but still ill-suited) to science and social science disciplines as deplorable and counter-initiative for the arts. For others however, the relationships are ones of potential evolution and critique for both art and Higher Education. Irit Rogoff, a European-based arts and visual culture writer, for example, writes of a wider educational turn within art. This both sits in critical relation to the university, critiquing it from within, whilst in other forms rejects institutionalisation through the creation of alternative forms of arts education.[72] Art writer Danny Butt, Associate Director (Research) of Victoria College of the Arts (now affiliated with University of Melbourne, but retaining its name at least) explores the potential of artistic research to help evolve 'future academies.' He sees practice-based research as a source of potential critical evolution of today's challenging cultures and politics of Higher Education.[73] Whilst, Montreal based SenseLab, led by Erin Manning and Brian Massumi, experiments with practices of research-creation to evolve 'alter economies' of the relationship between research and creative practices.[74] They are specifically concerned to critique the 'ensnarement' of relations between research and creative practice by the creative economy, and the rendering of the potential of these relations as mere 'institutional operators' on behalf of the neoliberal institutions of art and the university.'[75] The specific ways that tensions around the research turn and the politics and cultures of Higher Education play out for those located within the discipline of geography are explored in more depth across the chapters of this volume. It would be a mistake, however, to reduce the current research turn solely to those relations that have emerged through the explicit relocation of art schools in the context of Higher Education; there is also the emergence of what could be called 'research aesthetics.'

Research aesthetics is a term I want to use to denote those artistic practices which are conducted in the context of specific forms of research, and the ways these forms of research manifest within the works and exhibitions that are produced. In other words, I am concerned with artists using research practices – from ethnography to archival work – as a medium, with artists who use their research materials as the very literal material from which to make art (for example, the proliferation of walls of journal articles, books, mind-maps, data sets, and so on as installation environments). Research aesthetics is also concerned with those practitioners who enrol audiences as co-researchers within their art and its presentation within exhibitions.

To expand. There has been a notable trend within art of the late twentieth and early twenty-first century for artists to adopt and adapt the figure and practices of the researcher. This includes the site-specific artist who assembles in-depth research on a specific location – its pasts and presents – as well as the artist emerging as ethnographer, as historian, as archivist.[76] Oftentimes, for the art theorists who ascribe these roles to the artist they do so in dialogue with academic traditions of these practices. Hal Foster, in his two essays 'Artist as Ethnographer' and 'Artist as Archivist' offers contrasting views of the potential and practices of artists as researchers.[77] On the one hand, the artist as ethnographer emerges, damningly, as a practitioner fetishising complex research methods, keen to grasp the authority offered by the posture 'ethnographer,' whilst seemingly unaware of the controversial politics of modern ethnographic practices. On the other hand, the artist as archivist adopts and adapts established research methods to critical ends, extending and critiquing these methods as they do so. Foster's essays, read as a pair, suggest the artist valued as a 'good' or a 'bad' researcher. Artists are also turning their research materials, from data sets to literature reviews, as well as mind-maps and various paraphernalia of research in the field and the lab, into materials for constructing installations. This finds form, for example, in artists installing libraries or reading rooms within their exhibitions, sort of spatialised literature reviews, with copies of books, print-outs of papers, inspiring quotes selected and written on walls, as well as 'thought' diagrams of core ideas and organisational structures.[78] These modes of display demand new understandings of audience – in some cases as a co-researcher, or as subject to information overload. Claire Bishop conjures images of an audience lost in drifts of material, where the experience of research is akin to 'internet search,' fostering in the audience 'quick skimming, elliptic reading, browsing.' Here the volume of material is understood to offer a kind of 'logic' of research, evidence of its occurrence, a mark of authority, or even just for ornamental effect.[79]

It is clear that there are myriad instantiations of 'research' within the art world's research turn. As with the diversity of creative geographies however, it is equally clear that what this broad body of work tends to have in common is a desire for greater critical discussion and debate. On observing the growth of practice-based PhDs and their accompanying critique, Elkins notes while he has no worries over future student numbers, he does have concerns over the criticality of the debate.[80] For others, critical accounts have been stunted by too much focus on naming debates, or cataloguing existing work.[81] Elsewhere, debates about the form of research seem to get hamstrung by models of research that seem to owe much to natural sciences and asserting by contrast, tacit embodied knowledge through practice. A further common refrain suggests too little attention has been given to the practices and experiences of doing practice-based research.[82] Arguably, this need for criticality only gets more acute, when practitioners turn to other disciplines as a home for their practice, where, whether geography, sociology or archaeology, there is a chance to understand how differing practices of research are emerging together.[83] It is in this space that this volume exists, experimenting with a geohumanities approach to these comings together.

A GeoHumanities approach?

Our current understandings of these creative practice and geographical research relations is suffering, ironically from what could be caricatured as a forgetting both of the geo and of the humanities. What do I mean by this? I mean that while cultural geographers have long recognised the co-constitutive nature of creative practices (i.e. they make worlds) and have appreciated that these world-making possibilities are premised on the interaction of the geographies within these work's 'frames,' with those of their production, consumption, and circulation, they have been rather lax in applying these understandings to their own creative practices.[84] In other words, much literature focusses on what these creative practices do – engage us with embodied experiences of the world, offer forms of felt knowledge, enable us to listen to and engage with overlooked perspectives, help us reach new audiences – to name but a few ideas, but what the literature does not do in any concerted way is explore the role of the geographies of these practices. In short, appreciate how the 'wheres' of their production and consumption are vital to the how's, why's, and what's of the work. As such, ironically, some of the key lessons cultural geography might offer these approaches – their geographies – have gone missing. Further, accounts of creative practice and research relations within geography often seem to suffer not only from missing geographies but forgotten humanities. In other words, geographers have largely failed to draw on the rich humanities understandings of creative production, consumption, and circulation. These missings and forgettings are often mirrored within creative practice-based research. Not only have the geographies of these practices largely been overlooked, perhaps understandably, but perhaps more confusingly, little space has been made for humanities-based understandings of the research process. For some reason, discussions of creative practice-based research seem haunted by modes of research that hail from science and social science. The discussions of rigour, of critique, and of scalability seem ghosted by positivism, rather than taking account of those more humanities-based research practices that assert the individual and subjective, that recognise the liveliness and agency of materials, and that naturally valorise modes of the empirical other than the extractive.

The GeoHumanities experiment of this volume attempts to combine perspectives from geography and humanities, as well as drawing on bodies of work where this is clearly already the case (e.g. historical geographies and cultural geography), to offer a critical account of the myriad comings together of creative practice and geographical research. In doing so I am inspired by three bodies of work. Rather than attempt to summarise each of these vast and rich fields here, I want instead to draw out the main ways they shape what follows, while leaving the detail to the chapters themselves.

The first body of work concerns art histories and theories, and even some sociologies of art. I initially felt anxious about the scale and scope of this text – with its focus on something akin to a movement or medium, rather than an elaboration on a substantive geographic theme or single practitioner, as is more normal for geographic texts on art. Yet a core inspiration has been those art histories and art theories that explore the emergence, evolution, and impact of a particular mode of art making. Of relevance here would be Claire Bishop's *Artificial Hells*, which explores art that involves participation, or Miwon Kwon's *One Place After Another* which is focussed on site-specific artwork.[85] As in these volumes, I seek to work across a diffuse mode of practice –here creative practice and research relations – through querying the critical terms of artistic production, consumption, and circulation. The risk is that this reads more like a 'review,' but I would urge those who end up feeling this way to appreciate the value and importance of these accounts of a medium or field, and appreciate this as a form of analysis in its own right. As well as inspiring the form of this volume, ideas across art history and theory, principally under the influence of the spatial turn, also offered finer-grain critical contexts, including discussions of the studio and the site, as well as the shifting modes of art's circulation and audiencing that have emerged in the last decades.

A second inspiration for this text has been the geographical histories of science and geographical knowledge. The 'taking place' of knowledge, and an appreciation of both 'science in situ' (for this is mostly histories of science) as well 'science in motion' has been crucial for my attempts to plot the complex relations of the production, consumption, and circulation of knowledge.[86] This is an expansive literature that moves across times, spaces, and scales, in a consideration not only of those predictable sites of the lab and field, but also of the sites of museums, journals, lecture theatres and conferences, pubs and clubs, dining and drawing rooms.[87] These are both expansive accounts of learned societies, as well as appreciations of small stories of everyday spaces and thoroughly ordinary practices. For, as Dewsbury and Naylor put it, 'everyday actions make up the grander facades of institutional agendas, empirical projects and disciplinary schools of thought.'[88] Furthermore, these are studies that remind us of the importance of sites of the reception of knowledge or audiencing. As Innes Keighren and others make clear, there are multiple intersecting geographies at stake in understanding how the geographies of the book and the wider cultures of print, presentation, and reception shape science.[89] This takes in the sites of production as well as the circulatory networks through which print moves, the situated appraisal of texts, and the ways audiences encounter and interpret these outputs, and the geographies of who has the resources to do so.[90] Interestingly, these discussions have yet to engage in any depth with the histories of creative practices as forms of geographical knowledge production.[91] Further, rarely might such histories get connected with those diverse cultural, social, and economic geographies that attend to the geographies of the creative economy, from discussions of the studio to inspiration in the field, or the making of guitars, surfboards, or clothes in both creative clusters as well as commodity chains strung across the world.[92]

A third set of inspirations comes from a diffuse set of discussions that might be grouped by way of its critical reflection on disciplinary relations, whether termed inter, trans, cross, or post. Such forms are emerging– at least rhetorically – as one of today's most pervasive modes of knowledge making. Their ubiquity is met variously with delight, scepticism, and unappealing visions of a landscape overpopulated with homogenised interdisciplinary research centres.[93] Once described as the 'most seriously under-thought critical, pedagogic and institutional concept in the modern academy,' interdisciplinarity and its relations – intra, trans, multi, and cross – have increasingly become objects of interest for researchers.[94]

Geographers have long positioned their subject as one of breaking down disciplinary boundaries through crossings, pairings, intersections, interventions, and the like.[95] Disciplinary conferences and plenaries have examined inter and intra disciplinary relations as a way into wider considerations of geography's knowledge domain, history, and prospects.[96] In these various venues there has been concern and celebration over geography's intradisciplinary nature (parsing of its disciplinary wholism, speculation on how best to fulfil its transdisciplinary potential with respect to the environment or globalisation, and specification of its intellectual genealogies and the diverse research ecologies that foster interdisciplinarity).[97] What is more, with the wider spatial turn, geographers' own extra-disciplinarity has been lauded, celebrating its embrace of a wide range of different fields of study.[98]

Geography, of course, has no monopoly on such ways of working or disciplinary caricatures. Artists and other creative practitioners too have long identified with working across, and beyond, disciplines. The figure of the interdisciplinary artist has long been seen as a disruptive figure working across or 'in the midst' of multiple practices such as sculpture and dance, or installation and writing, putting at risk what is understood by these mediums.[99] Whilst the transdisciplinary artist has become a globally recognised figure, working often on 'wicked' problems, with or alongside climate scientists, engineers, and so on.[100] Cast in other terms, but no less relevant, are the myriad practitioners who think of their processes as a form of nomadism, or gleaning, amassing material from a diverse range of knowledge domains to address the problem at hand. A constant refrain when reflecting on these practices is a seemingly fine line between what might constitute intellectual dilettantism, as set against what might be conceived of as a drawing from across disciplines in a productive manner.

Across this rich body of work on disciplines and how we practice them, the prefix is often a site of considerable preoccupation – inter, trans, cross, post, and so on – as well as ongoing considerations of various metaphors of coming together or intersecting: bridging, hybridity, transversality, and so on. Tending to adopt the terms of those doing this work, this volume steers away from attempting to theorise these variations in favour of attending to practice. What it does take forward however, is the emerging body of work that explores the place of disciplines in knowledge making practices by way of empirical accounts of projects-in-practice. Examining how these forms of research are accomplished, how they respond to and constitute new research objects, and how they might cultivate new subjectivities and skills, these accounts offer both accounts of individual, but more often team- or group-based practices.[101] Of note here is Mike Pearson's (a site-specific performance practitioner) and Michael Shanks' (an archaeologist) agenda setting, *Theatre/Archaeology* published in 2002. Almost twenty years on, this co-authored text, which cuts together textual and image-based accounts from 'the convergence of two biographies/projects/ discourses,' feels extraordinarily prescient in its thinking together and apart of the intersections of creative practice and archaeological working.[102]

These three forms of inspiration weave through the eight chapters which follow. At times, I fear the frustration I often felt in trying to bring these per- spectives together might be writ too clearly on the page. For this weaving is far from evenly spaced – and I was haunted by those worries I was encountering on the page. There are densities of discussion where one perspective takes over, else- where the threads seem thin, rich in variety but light on detail. This variety, and my lack of surety at its success, is one reason why I imagine this as an experiment with a geohumanities approach – how this works exactly and what this resolves into is considered further in the conclusion.

Researching and writing research?

The research that forms the basis of this book has an uneasy origin. How to research research became a preoccupying question that concerned not just the methods used and their ethics, but also the validity of studying my own and my colleagues' research practices. It was also a question of what it meant to try to write about them in a way that did not seek to fix them through definitive tra- jectories or canonise certain forms of practice or ways of working. The project had oblique origins. It congealed out of needs posed by the growing force of creative practice-based ways of working within my various research projects on rubbish, place, environmental change, and the underground, and my own wider experiences of my own and others' struggles to make sense of these practices. As I experienced sticking points, and witnessed others, in turn, battling them, I turned to arts to find their solutions, and found a similarly conflicted and chal- lenged field. I went looking for inspiration (and justification) for researching the research culture I was embedded within (worrying that rather than desperately seeking guidance if not solutions, it might seem lazy or somehow self-absorbed), and found it within the diverse fields of studies of knowledge production and in particular, the anthropological accounts of disciplines and knowledge more widely, and the historical geographies of geographical knowledge.

'Doing anthropology among the disciplines' has become increasingly com- mon, not least under the influence of the social studies of science's laboratory ethnographies.[103] Further, academics are increasingly driven to reflect auto- ethnographically on their evolving practices of interdisciplinarity and collabora- tion, as well as elements such as ethics, participation, and public engagement.[104] This sits alongside a somewhat different vein of work that addresses the lived experiences of academic life, whether in terms of well-being concerns, issues around equality and diversity, or the challenges of doing writing or fieldwork.[105]

While 'doing ethnography at home,' indeed of one's own 'house,' might be increasingly common, intimate involvement in one's own research worlds is both rich resource and at times a liability. Like many academics studying their own field, I grappled with concerns over the legitimacy of data, drawing as I did on the 'informal gleanings of everyday insider experience' whilst worrying over how to 'remain systematically and critically attentive to the social conventions of the field communities' that are one's own.[106] I drew much inspiration from feminist assertions of the validity of the individual researcher, their subjectivity and the everyday as data.[107] As well as others writing on doing autoethnography on what is effectively life experience. As Dydia DeLyser and Paul Greenstein write of autoethnography, the

> retrospective power is one of the strengths of autoethnography: Like oral history, it links decades of life experience with deep introspective self-reflection, relying on the self- conscious self-examination of personal experiences and the insights of personal reflection over many years.[108]

Still, treating one's normal working practices – fieldwork, teaching, editing, reviewing, sitting on grant panels, event organisation, paper writing and so on – as field experiences works in complex ways. Issues included balancing the desire for research and feeling very foolish, being constantly aware of power relations, negotiating organisational and personal loyalty, fear of recrimination or reputational damage, and the need to probe sometimes difficult situations where entrenched views were hard to tackle in positive ways. Power relations are always tricky, but here are combined with relations with colleagues and friends that are cast through the lens of both a favour and the mutual benefits of open-exchange. The result was a persistent negotiation of anxiety, guardedness but also a heighted reflexivity as research-daily life became a site for the evolution of the social and academic conventions for a field community in which the researcher and her interlocutors are colleagues and friends. As my research interests became by necessity my research field, I shifted how I approached everything from grant writing to event organising, as well as reviewing and editorial work – making them, I think far better and more responsive as a result. Event series emerged from issues I was perceiving as shared across the field, collaborative practices gained further dimensions as a result of ongoing discussions, and reviews become sites for consciously writing through the issues of critique and evaluation I was encountering, almost as a form of dialogue with other editors and authors.[109]

Seeking to make sense of these tricky relations, I valued new ethnographic imaginaries emerging around paraethnography and allied concepts of lateral and collateral knowledge that premise ethnographic working on similarity rather than difference and otherness.[110] Anthropologist George Marcus, summing up these ideas, notes (in somewhat difficult language) the 'alter-native' forms that characterise practices where scholars work on spaces and knowledge economies with which they have affinities, and encounter subjects 'whose perspectives, curiosities and intellectual ambitions parallel the ethnographer's own.'[111] These are forms of ethnography shaped by 'intellectual co-creatorship' with 'knowledge counterparts' or 'epistemic partners.'[112] Necessarily then 'data extraction' is replaced with ideas of 'working speculations' and with the evolution of traded languages, 'shared processes and interpretations as well as objects.'[113] Worked through logically this approach also has impacts on how you write these ethnographies and the status they are accorded as knowledge forms. The influential writing culture debates of the late 1980s brought a political and epistemological crisis to the writing of ethnography as posturing of distanced reportage, pinning down the lives of others through the objective gaze.[114] The debates foregrounded instead, practices of ethnographic composition as a process of cultural representation that is contingent, historical, and contestable, a politics and poetics of cultural intervention that should also be appreciated as a 'troubled, experimental knowledge of self in jeopardy amongst others.'[115] The writing of paraethnographies requires this trajectory to be pushed even further. For these ethnographies premised on the in-the-midst might be understood as a speculative eventful writing that is thoroughly part of the becoming of the worlds it is working through.[116] This shifts the tenor of ethnographic writing, as Ingold has elaborated, away from a writing up of fixed accounts, towards a participation in the writing of tropes of possibility, anticipating, rather than singularly answering, questions, and working through and with (rather than on), ways of working that we are not yet quite able to clearly frame, pose and even, at times, to really make happen.[117] I have attempted to write a field-in-motion, an emerging and still evolving body of work, within which this text participates. As such writing tries to resist closure and finality, but instead seeks to invite moments of recognition, but hopefully encounters with difference on the part of the reader, potentially new possibilities and anticipatory practices that might aid in helping to co-produce the future of these forms of research.

In practice, my account emerged through an approach that combined autoethnography and paraethnography with methods drawn from historical geography and art history and theory, to evolve five key bodies of material. This includes the development of fourteen collaborative artistic research projects. Many of these are introduced throughout the volume. Hugely varied, these span from short projects of a few months, with much to learn but little in terms of 'outputs,' to an ongoing collaboration with artist Flora Parrott of some four-year duration at the time of writing, hugely formative for this text. The mediums are equally varied, taking in visual, installation, and community art, digital media work, curation, creative writing, and participatory practices. Another crucial body of material came from over fifty interviews with geographers and a range of creative practitioners, across human and physical geography, around the world and across career stages. I also conducted ethnographies of practice-based PhDs, including interviews with students and autoethnographic reflections on my own role as supervisor of ten and examiner of twelve of these PhDs. Another integral part of my methodology was a survey of "outputs" and extensive documentation of artistic research projects and creative research methods. Including attending to how these are produced, circulated, and consumed, within geography's 'scriptural economy' of journal articles, monographs, and spoken conference papers, as well as beyond it – in other journals and in catalogues, installations/ performances, artefacts, and experimental events. This involved my blending analytic perspectives from geography with those from art history and theory, an approach that has dominated much of my previous research.[118] A final crucial body of material is an ethnography of the evolution of the intersections between geographical research and art. This includes the development and maintenance of an infrastructure for creative practice and research relations. As well as interviews with key figures, textual analysis was used to explore the histories of journals, as well as diverse forms of research governance, including the evolution of practice-based PhDs within geography and the place of practice-based research within various national level evaluation frameworks. These include the UK's Research Excellence Framework (REF) and the associated Impact Agenda, the Excellence in Research for Australia framework (ERA), and the advent of 'Research Creation' as a funding category by the National Science Foundation of Canada. Inspired by the approaches of historical geographers studying their own disciplinary histories, a range of geographical writings also become akin to archival sources. These include monographs, journal articles, blogs, websites, Twitter, and Instagram feeds. Together they offer invaluable sources for exploring how these practices have proceeded, how creative practice and geographical research relations are being written of, and how these imaginations are circulating. Material was kept in field note-books, recorded and transcribed, using computer-based transcription service where appropriate. Participants were given the chance to remain anonymous, and all were sent the portions of the manuscript relevant to their work to comment on in advance of publication.

Going forward

Just as the geographies of science could never be reduced to a place-based mapping or classificatory system, neither can the geographies of creative practice and geographical research relations. Not only is the field, 'too rich, too fertile, too intricate' but of course as the with geographies of science, those of creative practice are not exhausted by accounts of location, but also take in mobilities, scales, networks, and so on.[119] There is no thought here of classification or of mapping, instead the suite of eight sites: field, studio, lab, community, residency, thesis, page, and exhibition is 'suggestive rather than comprehensive, illustrative rather than exhaustive, overlapping rather than separable, very much in motion rather than fixed.'[120] Each of the chapters addresses the book's three objectives: to explore some of the key sites of the coming together of geographical research and creative practice; to examine the understandings of art, geography, and research that emerge from the midst of these sites; and to investigate how our imaginations of these key sites are shifting and transforming in the context of these comings together. Whilst, as with science, the spaces and practices of the production, circulation, and consumption of these ways of working are thoroughly intertwined, the volume does evolve through three phases. The first trio of chapters – the field, the studio, and the lab – are iconic sites for the production of geographical research and art. The second phase focusses on a pair of sites – community and residency – addressing locations in which production and consumption are very firmly folded together. The final trio of chapters – thesis, page, and exhibition – takes up crucial locations for the circulation and consumption of these forms of knowledge making.

The volume closes with a set of reflections on the overarching lessons for the field of creative practice and research relations within and importantly, beyond geography. It returns to my three key objectives, and discusses them by way of themes that have emerged across the sites including questions of critique, concerns with process and product, with the varied 'place' of creative practice within the biography of a research project, and of course with epistemological questions. Again, the aim is not to police ideal practices but rather to work from the midst of these creative practices and research relations to query how it is that they enable reflection on and challenges to accepted research practices. The conclusion also reflects on this volume's geohumanities experiment. It probes those elements that from my perspective worked and those that were a real struggle. Not least, the tensions between different forms of empirical material, and the varied understanding of the empirical that sometime seemed to underpin them. The conclusion will also look up and out from geography as a disciplinary location to reflect more widely on the value of this volume's findings for other disciplines currently in the midst of a growth of relations between creative practice and research. This is an exciting and evolving field, but one which like all expanding fields of art before it, and evolving modes of knowledge production requires new modes of production and evolving forms of consumption and of critique. It is the possible forms of these, their possibilities and, importantly their potential, that this volume seeks to address.

Notes

1 See for example: Marston and De Leeuw, "Creativity and Geography"; De Leeuw et al. "Going Unscripted"; Crouch, "Choreo-graphic Figures."
2 Elkins, "The New PhD"; Cazeaux, *Art, Research, Philosophy*; Manning and Massumi, *Thought in the Act*.
3 See for example summaries in Crang, "After a Fashion"; Marston and De Leeuw, "Creativity and Geography"; Madge, "On the Creative (re)turn in Geography"; Hawkins, "Dialogues and Doings."
4 See for example; Crouch, *Flirting with space*; Nagar, "Storytelling and co-authorship"; Garrett, "Videographic Geographies"; McCormack, *Refrains*.
5 See for example discussions in Hawkins, *Creativity*.
6 I have attempted to indicate the global spread of work through the citations chosen, as well as the websites referenced, for, of course, much of this work, as Chapter 7

on the page discusses does not easily have a journal or book-based presence. In terms of the large research centres, a key example in Europe is the work on research creation being done at Université Grenoble Alpes, http://maisondelacreation. univ-grenoble-alpes.fr/en/arts-in-the-alps/spring-school-2017/arts-in-the-alps-spring-school-2017--164258.kjsp; another is the Mobility and Humanities, Centre for Advanced Studies at University of Padua. Two key Asian examples are Konkuk University's 'Academy of Mobility Humanities,' http://www.mobilityhumanities. org/main.html?lang=EN (accessed 23/1/2020); and the National Taiwan University Research Centre for Future Earth (accessed 23/1/2020), https://www.facebook.com/ NTURCFE/?pageid=239297520273597&ftentidentifier=407711730098841&padding=0

7 See for example, Cochrane and Russell, "Art and Archaeology"; Back, "Live Sociology"; Harman, *Seeing Politics.*

8 Barrett and Bolt, *Practice as Research;* Cazeaux, *Art, Research, Philosophy.*

9 Butt, *Artistic Research in the Future Academy;* Cazeaux, *Art, Research, Philosophy.*

10 Foster, "The Artist as Ethnographer?"; Foster, "Archival Impulse"; Bishop, "Venice Biennale."

11 Butt, *Future Academy;* Rogoff, "Turning."

12 Manning and Massumi, *Thought in the Act.*

13 Cazeaux, *Art, Philosophy, Research;* Elkins, "The New PhD."

14 Tolia-Kelly, "The Geographies of Cultural Geography"; McLean, "Hos in the Garden"; Last, "Experimental Geographies."

15 Manning and Massumi, *Thought in the Act.*

16 Lesage, "Who's Afraid of Artistic Research?"

17 Cazeaux, *Art, Philosophy, Research;* Elkins, "The New PhD."

18 Bishop, *Installation Art;* Bishop, *Artificial Hells;* Kwon, *One Place After Another.*

19 Mayhew, "Materialist Hermeneutics."

20 See for example the collected essays in Boyd and Edwardes, *Non-Representational Theory and the Creative Arts.*

21 Dixon et al, "Editorial."

22 Dear et al., *GeoHumanities;* Daniels et al., *Envisioning Landscapes, Making Worlds.*

23 Hawkins, *For Creative Geographies.*

24 Livingstone, *Putting Science in its Place;* Agnew and Livingstone, *Handbook of Geographical Knowledge.*

25 Daniels, *Fields of Vision;* Rose, *Visual Methodologies;* Hawkins, *For Creative Geographies.*

26 Withers and Livingstone, "Thinking Geographically about 19th Century Science."

27 Discussed in Genosko, *Felix Guattari.*

28 Manning and Massumi, *Thought in the Act,* p. 88.

29 Ibid.

30 Greenhough, "Tales of an Island-Laboratory," p. 225.

31 Hawkins, *For Creative Geographies;* Mould, *Against Creativity.*

32 Osborne, "Against 'Creativity,'" p. 507; Mould, *Against Creativity.*

33 Osborne, "Against Creativity"; Gibson and Klocker, "Academic Publishing as 'creative' industry."

34 I borrow this term from Thomas Jellis's wonderful thesis 'Reclaiming experiment: geographies of experiment and experimental geographies,' and he, in turn, borrows from Brian Massumi, who is using it with respect to affect. Thesis at: https://ora.ox.ac. uk/objects/uuid:39de7269-7ddf-4aaa-a4a1-ae6ad9ed17bb (accessed 12/14/2019).

35 Hawkins, *Creativity;* Edensor et al., *Spaces of Vernacular Creativity.*

36 See for example, Jellis, *Reclaiming Experiment.*

37 Hawkins, *Creativity.*

38 Osborne, *Anywhere or Not at All.*

39 Hawkins, "Geography's Creative (re)turn."

40 Ibid.

41 Crang, "After a Fashion"; De Leeuw and Hawkins, "Critical Feminist Geographies"; Madge, "On the Creative (re)turn"; Hawkins, "Dialogues and Doings"; Last, "Experimental Geographies"; De Leeuw and Marston, "Creativity and Geography"; Boyd, "Non-Representational Geographies"; Sachs-Olsen, *Socially Engaged Art and the NeoLiberal City;* Pratt and Johnson, *Migration in Performance;* Cresswell, *Maxwell Street.* See also special sections of journals such as the 'cultural geographies in practice' section of *cultural geographies* and the 'practices and curations' section of *GeoHumanities.*

42 Madge, "Creative Re-turn"; Hawkins, *For Creative Geographies.*

43 Hawkins, *For Creative Geographies.*

44 See Chapter 6 for discussions of PhDs, books include Boyd, *Non-Representational Geographies;* Boyd and Edwardes, *Non-Representational Theory;* Magrane et al., *Geopoetics in Practice.*

45 See for example, Garrett, "Videography"; Brickell, "Participatory Video Drama Research"; Peterle, "Comics and Maps."

46 Pink, *Doing Visual Ethnography.*

47 McCormack, "Drawing Out Lines."

48 Wylie and Webster, "Eye-opener"; McCormack, *Refrains;* Dewsbury, "Seven Injunctions."

49 See for example McCormack, *Refrains.*

50 See essays in Boyd and Edwardes, *Non-Representational.*

51 Straughan, "A Touching Experiment"; McCormack, *Refrains;* Hawkins, "Creative Geographic Methods."

52 Foster and Lorimer, "Some Reflections"; Wylie and Webster, "Eye-Opener."

53 Patchett and Mann, "Five Advantages of Skill."

54 Barnett, "Cultural Twists and Turns."

55 See for example, Tolia-Kelly, "Fear in Paradise"; Askins and Pain, "Contact Zones"; Brickell, "Participatory Video Drama"; Gorman-Murray and Brickell, "Over the Ditch"; Harel-Shalev et al., "Drawing (on) Women's Military Experiences"; Nagar, "Storytelling and Co-authorship"; Hawkins, "Doing Gender and the GeoHumanities."

56 De Leeuw et al., "Going Unscripted."

57 Ibid., p. 157.

58 Brice, "Situating Skill"; Rogers, "Advancing the Geographies of Performing Arts."

59 Tolia-Kelly, "The geographies of cultural geography"; McLean, "Hos in the Garden."

60 https://www.documenta.de/en/retrospective/documenta_13 (last accessed 12/2/2020).

61 https://universes.art/en/venice-biennale/2019/research (last accessed 12/2/2020).

62 http://theoria.art-zoo.com/picasso-speaks-pablo-picasso/ (last accessed 12/2/2020).

63 Hogarth's satirical print can be seen at https://research.britishmuseum.org/research/collection_online/collection_object_details.aspx?objectId=1433960&page=1&partId=1&peoA=120911-2-60&people=120911 (last accessed 12/2/2020).

64 Cazeaux, *Art, Research, Philosophy;* Slager, "Nameless Science."

65 Cazeaux, *Art, Research, Philosophy;* Butt, *Future Academy.*

66 Hockey, "United Kingdom."

67 See for example, Klein, "What is Artistic Research?"

68 Elkins, *Artists with PhDs;* Butt, *Future Academy;* Slager, "Nameless Science."

69 Elkins, "The New PhD."

70 Relyea, *Your Everyday Art World;* Raunig, *Factories of Knowledge.*

71 Relyea, *Your Everyday Art World.*

72 Rogoff, "Turning."

73 Butt, *Future Academy.*

74 Manning and Massumi, *Thought in the Act.*

75 Ibid, p. 88.

76 Foster, "The Artist as Ethnographer"; Foster, "Archival Impulse."

77 Foster, "The Artist as Ethnographer."

78 Recent examples of this practice in very different contexts include the Oslo Architecture Triennale's library, the Triennale's chief curator was geographer Cecilie Sachs-Olsen; the 'Extended Studio,' in display in Tate Modern's, London, 2019–2020 Olafur Eliasson's retrospective "In Real Life," and the lab and research room accessible to the public during Tomas Saraceno's "On Air" 2018 retrospective at the Palais De Tokyo, Paris.

79 Bishop, "Information Overload."

80 Elkins, "The New PhD."

81 Lesage, "Who's Afraid of Artistic Research?"

82 Ibid; Cazeaux, *Art, Research, Philosophy*.

83 See note 3.

84 Rose, *Visual Methodologies*; Hawkins, *For Creative Geographies*.

85 Bishop, *Artificial Hells*; Kwon, *One Place after Another*.

86 Livingstone, *Putting Science in its Place*; Keighren, "Geographies of the Book"; Ogborn and Withers, *Geographies of the Book*.

87 Livingstone and Withers, *Geographies of Nineteenth-Century Science*.

88 Dewsbury and Naylor, "Practising Geographical Knowledge," p. 254.

89 Keighren, "Geographies of the Book"; Ogborn and Withers, *Geographies of Nineteenth-Century Science;* Keighren et al., *Bringing Geography to Book*.

90 Livingstone, "Science, Text and Space."

91 Although see Daniels, "Art Studio."

92 Warren and Gibson, *Surfing Places*; Leyshon, "The Software Slump"; Brace and Johns-Putra, "Recovering Inspiration."

93 Barry and Born, *Interdisciplinarity*.

94 Liu, "The Power of Formalism."

95 Massey, "Space-time"; Lane, "Constructive Comments"; Clifford, "Physical Geography"; Bracken and Oughton, "What Do You Mean"; Castree et al., "Boundary Crossings"; Baerwald, "Prospects."

96 Mona Domosh, President of the AAG 2015 selected 'radical intra-disciplinarity' as her Presidential Plenary theme at the 2015 meeting in Chicago.

97 Donaldson et al., "Mess"; Buller, "The Lively Process"; Wainwright et al., "Geographers Out of Place"; Dixon et al., "Sublimity, Formalism."

98 Thrift, "Future"; Warf and Arias, *The Spatial Turn*.

99 Condee, "The Interdisciplinary Turn."

100 See for example, Hediger and Scott, *Recomposing Art and Science*; Scott, *Processes of Inquiry*.

101 See for example, Foster and Lorimer, "Some Reflections"; Driver et al., *Landings*; Callard and Fitzgerald, *Rethinking Interdisciplinarity*.

102 Pearson and Shanks; *Theatre/Archaeology*, p. 2.

103 Pearson and Shanks; *Theatre/Archaeology*; Mol, *Body Multiple*; Barry and Born, *Interdisciplinarity*.

104 Callard and Fitzgerald, *Rethinking Interdisciplinarity*; Kindon et al. *Participatory Actin Research*; Pain, "Impact."

105 Ahmed, *On Being Included*; DeLyser and Hawkins, "Writing Creatively"; Kitchin, "Engaging Publics."

106 Lederman, "The Perils of Working at Home."

107 Ahmed, *On Being Included*.

108 DeLyser and Greenstein, "The Devotions of Restoration."

109 Lederman, "The Perils of Working at Home"; Bradberg, "Being There."

110 Konrad, *Collaborators Collaborating*.

111 Holmes and Marcus, "Cultures of Expertise"; Holmes and Marcus, "Fast Capitalism."

112 Marcus, "Ethnography Two Decades"; Konrad, *Collaborators Collaborating;* Holmes and Marcus, "Cultures of Expertise"; Holmes and Marcus, "Fast Capitalism."

113 Marcus, "Ethnography Two Decades"; Holmes and Marcus, "Cultures of Expertise."

114 Clifford and Marcus, 'Writing Culture."
115 Ibid.
116 Ingold, *Being Alive*, p. 15.
117 Ingold, *Anthropology is Not Ethnography*.
118 See for example Hawkins, *For Creative Geographies*.
119 Withers and Livingstone, "Thinking Geographically about 19th Century Science."
120 Ibid., p. 12.

1

FIELD

It was a wet day in June. I had forgotten my waterproof. We drove through a very muddy field, across some open land and parked up near some trees and a wooden fence. We carried our kit through the woods, a short walk up to the cave site (Figure 1.1). Artist Flora and I were visiting Gully Cave, Somerset (SW England), a site that has been the subject of over a decade of excavations by a team of physical geographers in our department of Geography at Royal Holloway. They have been finding, logging, and identifying Pleistocene faunal remains, using these as proxies to explore past climates and to think forward into the future.

There was a generator, a tarpaulin, an archive section. After a tour, we got down into the cave, at that point some ten feet or so, and started scraping, trowelling at the earth, putting it into buckets. We suspected we had been given an area in which no one really expected to find anything. It is very calming, patient work. In the cave there are plumb lines, hanging threads, there are trowels, buckets, leaning mats. There is a high degree of precision, finds are mapped and plotted in three dimensions, logging their place in space and so in time. But there is also a significant amount of making do and getting by, a makeshift tarpaulin covers the site from the weather, it was being rearranged when we were there, it is somewhat ad hoc, but it clearly works. We reflect on the temporalities of research, digging for over a decade, and the place of the imagination as the team uses the finds to story the past of this place: flood waters choke the cave, the roof falls in, inhabitants, including bears, are buried. Some struck flint flakes, burnt bones and charcoal indicate hominin presence.[1]

Every year, after a few weeks at the cave, Prof Danielle Schreve and her team bring sediment back to the lab to sift and sort for tiny bones and teeth that might have been missed and might offer the clue to something's life, behaviours, and diet in the cave. It is not known how far down they can dig before they meet the floor of the cave – will each year bring an end to the excavation?

Six months later, the Bussey Building, Peckham.

FIGURE 1.1 Gully Cave, Somerset, England. (Image, author's own)

FIGURE 1.2 *These Pits and Abysses*, The Bussey Building Peckham, Flora Parrott (2016). (Image, author's own)

FIGURE 1.3   *These Pits and Abysses*, The Bussey Building Peckham, Flora Parrott (2016). (Image, author's own)

Danielle, head of the Gully Cave dig stands in the midst of an installation Flora has made. Elements of the field site have been transposed into the gallery space: the grid from the cave is marked out on the floor in hatched chalk and hazelnuts (Figure 1.2). Finds are marked with sculptural forms and tags hanging from the ceiling, located in imagined time-space. Danielle stands and talks in the midst of the grid of finds. She describes the histories, the time frames, the animals involved. She contours the spaces of the past in front of us. Whilst she never refers directly to the space or sculptural objects she stands within, the objects and her words weave together, inviting us to reflect on the imagined and material field practices that summon past worlds from earth and bones (Figure 1.3).

There are two other 'fields' in the installation space. One, occupied by political geographer Rachael Squire forms a kind of fabric capsule, streams of printed bubbles rippling as the air moves across the space. Standing in this undersea space, Rachael offers an account of her research on SEALAB, a US Naval experiment in undersea habitats developed during the 1960s. In the final under-surface field space Flora has created, I sit as if in a void. A series of sculptural fans printed with underground images surround me. There is a black fabric backdrop onto which is projected a live stream feed from the camera hanging from the ceiling pointed downwards at my head. It looks like I am sitting in the bottom of a hole (Figure 1.4). I read a text that draws together a sequence of six different underground imaginations.[2]

Accounts of field and gallery work with Flora Parrott

It seems appropriate that we begin in the field. The field has long been a primary site for the 'doing' of geography; indeed it has been understood as the site of

FIGURE 1.4 *These Pits and Abysses,* The Bussey Building Peckham, Flora Parrott (2016). (Image, author's own)

'real' geography and the locus of 'becoming of the geographer.' Yet it has also, especially in its inheritances from exploration, become lambasted for its masculinism, colonialism, and romanticism.[3] Felix Driver however, noted that despite its importance to the discipline, geographers had not really attended much to the field.[4] He urged us to pay more attention to the 'materiality of the field, the contingency of encounters within it, and the embodied practices of fieldworkers themselves.'[5] In the wake of this call, other more complex cultures and practices of fieldwork began to be built for geography.[6] Meanwhile, within the art world, the field has become a site of interest. We see a growth in artistic practices conducted through fieldwork and exploration, such as the Unknown Field Division's exploration practice or Cape Farewell's trips to the Arctic and the Amazon with artists and scientists.[7] We also find the field being recreated in gallery spaces, such as in Flora's work *These Pits and Abysses,* or in Mark Dion's archaeological digs.[8] Other artists are engaging with the material culture of the field, producing field-kits and guides as sculptural practice and textual form to induce audiences into field imaginations or encourage them into the field themselves. We might think here of the collaboration between geographer-poet Eric Magrane and writer Chris Cokinos which resulted in *The Field Guide to the Sonoran Desert.*[9] We even see the writing of field-reports and the creation of field-guides as increasingly popular forms of catalogues and gallery-guides.[10]

If both geography and art have a growing interest in the field, this chapter sits in dialogue with this interest through its exploration of the 'field' as a site for the unfolding of creative practice and geographical research relations. It does so through the lens offered by the folded modalities of fieldwork that occurred over the course of a Leverhulme Trust-funded artist residency conducted by Flora Parrott in the Geography Department at Royal Holloway, University of London.[11] During this residency we brought together our interests in the underground as a site of scientific practices, embodied experiences, and imaginative potential. In retrospect the residency unfolded as a series of ongoing field and gallery-based experiments (although I use this latter word with caution), of which I will offer a further two examples in this chapter, others appear in later chapters. The visit to Gully Cave, detailed at the start of this chapter, offers a microcosm of these fieldwork forms. First, it took Danielle's fieldwork in the cave as an object of research; second, in doing so it constituted fieldwork for our own investigation of the underground; third, I was entwining (not without challenges), fieldwork on the underground with fieldwork on what artists do 'in the field.' Finally, Flora created/recreated 'field-spaces' through her installations encouraging us to encounter these spaces anew. These folded fields constituted a productive sequence of field experiences that stimulated at once both intense points of recognition, where the field and what I was doing within it clearly slotted into accepted geographical practices and imaginations of fieldwork, but also moments where I felt very unsettled. It is in the midst of these points of recognition and difference that this chapter's discussion of creative fieldworkings emerges.

Situating my sequence of field experiences in the context of other accounts of creative practice-based fieldwork offered by others in written texts and during interviews, this chapter explores the imaginaries of the field and fieldwork that emerge. In exploring these myriad creative fieldworkings, this chapter recognises the field as a site where we both reflect on and shape how we actually do geography: methodologically, epistemologically, conceptually, and ethically.[12] The field is a key site where geographical problems are defined, analysed, and actually acted on, but it is also a 'shared space,' that as Louise Bracken and Emma Mawdsley suggest, crosses the sub-boundaries of Geography as a discipline, helping to shape its contours.[13] Both geographers and wider science studies scholars have recognised, as Felix Driver puts it that 'the theme of fieldwork raises questions about the boundaries between different kinds of knowledge.'[14] Indeed, the field repeatedly emerges as a site of encounter, an unsettling space that puts into question disciplinary identities and methodological strategies.[15] As such, this chapter joins those recent explorations of fieldwork that enjoin us to attend to the nature of fieldworkings and to 'the very acts of boundary-work which define our field– spaces.'[16] The fieldwork Flora and I did with Danielle drew a lens onto intersections of human geography, physical geography, and art.[17] This focus is deepened through two further accounts of being in the field with Flora – in another cave and on a glacier. Ahead of this however, I want to reflect on other instances of how creative practices have begun to reshape wider disciplinary imaginations of the field.

Reimagining the field

Geography has long been engaged in a project to reimagine the field. I would include within this project diverse work on the histories of geography, from post-colonial critiques of exploration to discussions of the bodies considered proper for fieldwork; debates over whose stories – and thus forms of fieldwork – are allowed into the 'canon'; and the kinds of techniques that are valued as opposed to those that are dismissed.[18] I would also include the wealth of scholarship that addresses how new concepts often demand new field practices. This could, of course, cover almost any form of methodological literature, including embodied, emotional, affective, sensory, and practice-based approaches to the world.[19] Oftentimes the epistemological questions these ideas raise dispose us towards not only new methods, but also demand new ways of thinking about the space-times of research itself. Working in these ways requires attention to how we enter and exit the field, how our sense of the field interacts with how we 'write-up' (or don't). In some cases, as will be explored further in what follows, we might need to extend imaginations of the field that are predicated on the extraction of data which is brought back to the lab, study (or studio), with other ways of thinking through our fieldworkings and our relationship with the spaces of the field.

In asking how the relationship between creative practices and geographical research might contribute to the ongoing project of reimagining the field, I draw inspiration from Anthropology's discussion of creative practices and ethnographic field practices. Two major anthropological figures are key here: George Marcus's reflections on site-specific and participatory art, and Tim Ingold's work on drawing and making.[20] Both find – albeit in different ways – great critical potential in creative practices for a reimagining of ethnography, anthropology's defining fieldwork practice. Teasing out key threads of their discussions helps set the scene for my own empirical reflections which follow.

George Marcus, together with James Clifford, was a key proponent of the writing culture debates of the 1990s.[21] They proposed an attention to the practice of writing ethnographies as a site at which to respond to the crisis of representation and the need for a critical account of the knowledge claims of modern Anthropology. At the heart of Marcus's more recent attention to artistic practice seems to lie a desire for alternative field cultures to those which, he claims, remain wedded to a Malowinoskian ethnographic *mise-en-scène*, despite the advancements made in ethnographic writing.[22] Marcus wishes to challenge the imagination of an anthropologist entering a field site in a far distant exotic location, embedding themselves in that locale, before extracting themselves and their data, and returning to 'write-up' their factual findings.[23] Marcus goes so far as to name the style, form, and imaginative ideal of ethnography this creates an 'aesthetics' of fieldwork, an aesthetics that here reads as somewhat Kantian in its formulation of aesthetics as 'ideal type' (beauty, sublime, or so on).[24] He is concerned to bring about a shift in fieldwork, which has remain untouched, he claims, by the writing culture debates, still reproducing dated 'technologies, aesthetics and power-knowledge' relations.[25] He finds in art new modes of field conduct; he writes, 'the kinds of research that some artists do are models that anthropologists can think within in articulating manifest changes in their own traditions of fieldwork.'[26] He finds particular inspiration in site-specific artistic practices, as well as so-called relational aesthetics (where, simply put, artworks configure social relations amongst individuals and groups, discussed further in Chapter 4 on community).[27] These models of art making seem to serve Marcus more as possible modes of field relations rather than as field practices per se. As his chapters, keynotes, and papers suggest, these artworks inspire new models of collaboration with participants, the possibility of ethnography as a form of intervention rather than as a mode of reproduction, and an attitude to 'outputs' that values impact at the site of study rather than distanced reportage.[28]

In contrast to Marcus's situation of creative practices as 'prompts to thought,' Ingold embeds creative practices in the doing of fieldwork and evolves a somewhat different sense of what such fieldworkings might do.[29] Ingold's practices are less those of contemporary art and more what could be considered 'foundational' creative practices of drawing and sketching. When combined with artisanal making – from basketry to kites – an attentiveness to embodied practice and materiality emerges.[30] Ingold, like Marcus, makes bold claims about what field practices of 'following materials, learning movements and drawing lines' might enable.[31] Through his work, drawing and dancing amongst other practices emerge as ways to 'enter into a world-in-formation in which things appear not as externally bounded objects, wrapped up in themselves and presenting only their outwards spaces for inspection, but as confluences of materials that have momentarily melded into recognisable forms.'[32] Skilled practices, he observes, are a response to 'moment-by-moment variations in the environmental conditions of their enactment.'[33] He finds in drawing, for example, a means to 'think-do' such that

> we do not first observe and then go onto describe a world that has already been made- that has settled into final forms of which we can give a full and objective account. Rather we join with things in the very presses of their formation and dissolution.[34]

This is a somewhat different aesthetics of fieldwork to that which Marcus describes, at the root of which is process, and which foregrounds 'pro-ductive' rather than 'ab-ductive' methodologies. As such, creative practices are enrolled within Ingold's wider formulation of pre-theoretical, practice knowledge (a knowing-how) as the basis on which all theory (a knowing what) rests. Drawing, weaving, dancing amongst other practices emerge as 'a gestural, rhythmic, uncertain negotiation between eyes, paper, pencil, hand, line, mind and heart.'[35] These creative practices become here shorthand for an epistemology that refuses a linear sequence of 'perception-judgement-action'; instead, Ingold writes, 'through the coupling of perception and action, the artist is drawn into the world, even as he or she draws it out in the gestures of description and the traces they yield.'[36]

I want to hold onto the worth anthropologists have found in creative practice for rethinking practices of fieldwork and researcher-field relations, as I consider how the varied intersections of research and creative practices might offer some contributions to geography's own project to reimagine the field. What follows explores two situated accounts of doing fieldwork, seeking to understand the diverse forms of creative fieldworkings that emerge from these experiences. We begin in another cave in Somerset, SW England.

Field 1: Cave: a sensory field?

"It was an odd sort of fieldwork." Was how the reflective section of the notes of one of my first fieldwork experiences with Flora began, this was to be an ongoing refrain. As my notes detail, it all felt quite familiar at first, perhaps even a little like a caricature. We got in a minibus, complete with crunchy Gore-Tex, loaded up with tents and kit, surely how many geography field trips begin? Eight of us – assorted artists, writers, and two geographers – a physical geographer and myself – some camping equipment and a three-year-old.

When we arrived at the site, we moved around the back of the minibus, donned all-in-one boiler suits over our wellingtons and set out. It was sticky in the suits, and colliding with various tree branches along the way it was clear I had no sense of the new dimensions of my head in its plastic helmet. We walked up the hill to the cave. This, we were told was sort of a training cave, no water, supposed to be easy, some 'nice' features (meaning a low ceiling and a corkscrew), and a clear end. Still we all seemed a little on edge. We came to the entrance; it was locked with a grill. Our guide, a member of the local caving club, explained that this was less to protect the public than to protect the bats who hibernated there. They had mostly, he explained, left now.

In we go. I was in a hurry, almost rude in my anxiety to get it over with, to know how bad it might be.

There are sections of crawling, a sort of dome-like space you could easily stand up in with some muted crystal geodes we take turns to peer at. There was the corkscrew – for a few seconds seemingly pinned with legs one way, head and shoulders the other, some sections of low roof. A slope we all sat on for a bit, the roof above us, torches shafting light around suggests this was shaped like a coffin – the cave was named after this feature. We turned off the head lamps and sat in the dark, exploring the effects of this much darkness. We were invited to climb down a ladder to visit the current 'end' of the cave. Some of us did. On the way back up we encountered a sole bat hanging from the ladder's anchor line, we carefully moved around it. A few people had been taking photographs throughout, not many though. Others had been making sound recordings of echoes, of breathing. Before leaving the large chamber we sat, poised at the top by the entrance, while someone wielded a camera around the space, using an iPhone app to take a series of images that would later be stitched together into a three-dimensional model of the space – rudimentary photogrammetry.

Later on that evening, we sat around hurricane lamps and the fire looking at a series of books about caving. We turned over pages considering photographs but mostly looking at the simple line drawings accompanied by often quite detailed written accounts of the routes, dotted with caving lexicon that made little sense to us. We compared the written account of coffin cave with our experiences, tried to map the two together, looking repeatedly at the simple line sketch. Later still, in my tent by torch light, I tried to make notes. Reading them back their focus is the embodied experience of caving, inflected by cultural geographical accounts of being underground. I reflected on the press of the rock, the experience of the dark, mused on the temporal experiences of being within such old spaces. I struggled to say much about the 'art' — fullness of this artistic fieldwork.

Account of fieldwork with Flora Parrott, in the Mendips, Somerset, 2016

The imaginary of fieldwork as a feat of masculine heroism hovered over this caving trip, as it did many of other research trips. This, at times, became personified in some of the guides we used, who initially started out by assuming that we desired a certain kind of experience of the underground, and in whom the underground inspired the telling of tales of extreme experiences and of near misses. The masculinism of the field, while roundly critiqued across geography retains, like the Malowinoskian *mise-en-scène* of ethnography, an enduring power.[37] Yet it has come in for significant critique as geographers follow Driver's and others' urging and attend to the embodied and practiced nature of fieldwork, to difference and the need to decolonialise field practices, challenging the centuries long image of the heroic explorer.[38] In this section I want to combine reflections on my fieldwork with Flora, with accounts of other forms of creative geographic fieldwork to probe the familiar and unfamiliar aspects of these field experiences that emerge.

The caving trip was one of several that I undertook with Flora and a group of seven artists and one physical geographer to the Mendips in Somerset. The trips were funded as part of Flora's Arts Council England Grant, Swallett.[39] This project, which overlapped in terms of theme and time-frame with our Leverhulme Artist-in-Residence Grant (discussed further in Chapter 6), was based on a series of field visits to cave sites and gallery-based working (discussed in Chapter 8) and work Flora did in the Royal Geographical Society (with Institute of British Geographers) archives. No outputs were promised, rather the activities were framed as open-ended occasions for thinking together at these sites. As the extract from my field notes make clear, this felt like an odd sort of fieldwork; indeed, I struggled to make sense of how this was fieldwork. These field visits did not really seem to be about data collection, not in ways I was used to. Nor was this a pilot visit following which we would reflect, evolve a research plan, and return. During much of the experience, as my fieldwork diary notes, I was far too caught up in being present and negotiating the tunnels and ledges to do much of what I thought of as 'data' collection. Yet at certain points of pause, we did begin to generate materials that would come to later inform art (sound recordings or images), and the reflections that we shared in the campsite that evening and afterwards together and apart at various events felt like 'data.'

Back in the camp, and thinking about the after-effects of being 'in-the-field,' it became clear that the embodied experience of the caves, the gaps between these experiences, and the line drawings that the books contained were points of inspiration for many. For others, the dark and what that did to our ways of knowing the landscape was a key preoccupation. In this our experiences resonated with those that have shaped wider geographical accounts of caving. Often these experiences are understood as bringing us to our bodies in particular ways.[40] They also resonated with wider accounts of doing fieldwork through the body, where the body is cast as a kind of sensing device.[41]

The figure of the fieldworker that emerged during these caving visits was a thoroughly sensory one. The work of being in the field became the work of becoming attuned to very different spaces, to the experiences of being underground, settling into the feel, sounds, and smells of the subterranean spaces and what they demanded of us. It was also a process of becoming aware of nerves calming, hands and feet becoming surer, and a realisation of what it meant to explore in the dark. Going caving is perhaps an extreme experience, but an attentiveness to fieldwork, as an embodied sensory practice, is a common imaginary to emerge across accounts of doing creative practice within geography. Capturing the spirit of many such accounts visual anthropologist Sarah Pink describes how, 'visual methods are mediating practices wherein mediation is a process that allows us to attain richer and fuller translations of bodily experience and materiality that are located, multi-textured, reflexive, sensory and polysemous.'[42] There is, of course, a lot to celebrate in Pink's description. Indeed, this understanding of creative methods as devices, as mediating practice through which we can better access had to get at 'data' but also 'translate' it for other audiences, is common across geographical accounts of using creative research methods in the field. Yet there was much that did not seem to really quite 'fit' with my experiences and my observations of the artists on the field trip. Whether this be the idea of creative practices as mediating access to something, or as having the aim of producing an account of these places.

For centuries, the value of creative practices for enabling different accounts of the field has been recognised. Indeed, the British Admiralty, reflecting on the fitting out and staffing of expeditionary ships in the eighteenth century, observed that artists might offer 'better accounts than those of words alone.'[43] Whilst, for humanistic geographers writing in the 1960s creative practices might offer 'nuggets of information' and 'pictures of places' that might otherwise be overlooked.[44] Yet of course, as historical scholars have demonstrated, making such 'truth' claims for art is far more complex than we might assume. Indeed, whilst the rise in *en-plein-air,* 'on the spot,' ways of working was predicated on the value of their practices of direct observation in the field, these were forms of art shaped by science, empire, and artistry.[45] For the art academy, the production of art as field-based 'authoritative knowledge about a distant place' was understood as a winter of the imagination.[46] Yet even at the time, it was clear that for many artists on board ship this was Tahiti or South America by way of the landscape aesthetics and practices of J.W.M Turner, Richard Wilson, or Claude.[47] The rich accounts of the era's evolution of ethnographic and botanical illustration as well as landscape paint-ing, detail the nuanced influence of aesthetic European forms on the apparently 'truthful' topographic images and scientific illustrations.[48] So, while the 'fields of vision' of such images might proffer 'packets of information,' these can be no more understood as a form of objective 'truth-telling,' than our current creative methods should be situated as solely part of 'know and tell' methodologies. In other words, that we value creative practices as primarily offering 'better' ways to research and present forms of knowledge – sensory experiences, emotions of affect – that might escape normative methodological practices.

To return to the cave. I was not convinced that the point of this particular artistic field experience was to generate or mediate accounts of otherwise hard to access things – the cave experience – to collect sonic, visual, or other forms of data on these caves for later use in artwork. Instead there seemed to be something else going on here. Reading across other accounts of creative practices in the field, it seemed that there was a similar ambiguity at work in them. Further, on the one hand creative practices were understood as offering access to other kinds of infor-mation about a place, or a community – human or non-human – but on the other hand, there was a resistance within the accounts of this being somehow a 'truth.'

To look at drawing practices for example. Helen Scalway, an artist-researcher on a large project 'Fashioning Diaspora Space,' details how her feelings on en-countering urban patterns translated into the kinds of marks she would make on the page. She writes evocatively of how the 'slice of pen against steel rule or the free-hand drawing of floral forms' constituted a bodily enactment of different paces and patterns of a place.[49] As she concludes, the performance of practices of hand drawing enacts and thus helps us come to know how different 'designed purposes (rational, traditional) developed in different cultures interact to produce the cosmopolitan city.'[50] Sage Brice reflects on her process of drawing cranes in the Huleh Wetlands in the Jordan Rift Valley as a practice of bodily opening to non-human others. She writes of the

> Process of becoming open (corporeally, social and politically) to the cranes and the complex histories of the nature reserve with the marks made on paper becoming the traces of this process of attunement — of learning and unlearning through a lived encounter.[51]

For those geographers interested in sonic methods, as Michael Gallagher writes, at-tending to sound requires a shared sense of vibrating bodily territories.[52] He ar-gues, with colleagues Johnathan Prior and Anja Kanngieser that, we need to develop broader sonic sensibilities, 'evolving an expanded concept of listening that concerns the responsiveness of bodies encountering sound- bodies of any and every kind in different ways and contexts.'[53] This includes, for example 'infrasonic vibrations that connect bodies across planetary distances, with the oceans, earth and atmosphere transmitting signals.' Such signals are only detectable to humans with specialised listening techniques, or through artistic work which helps us come to listen.[54] These accounts, like Ingold's, go back and forth with how creative practices might develop new ways of knowing, as well as collecting data that might inform later research or go towards the production of artefacts for later looking at or listening too by others. John Wylie, writing with artist Caterina Webster, observes of his experience of learning to paint and draw landscape that this is,

> on the face of it, to learn a new way of seeing, and in so doing to be drawn out, drawn in – drawn closer to visible objects, patterns and relations, to be aware of them anew – this is what I am trying to say. Looking at the canvas, and looking out; looking between them: these were exercises in closer attention, immersion and absorption. From what I have read and understood, this is almost an established narrative concerning the lived experience of drawing, in which drawing is understood in terms of intimacy, beckoning and revelation, such that when you draw, so you are drawn into the world.[55]

Throughout these accounts, including in my own reflections on using drawing methods, ideas such as 'drawing closer' 'listening better' suggests the ongoing association of these methods with offering access to 'data' otherwise hard to grasp. Yet something else hovers around the edges of these accounts, challenging the production logics of a 'richer,' 'fuller' account of sensory experience, or the logic of trying to 'access some previously unmapped marrow of the lived.'[56] As Wylie continues reflecting on his own practices;

> drawing is a lifeline; a mode of ours and the world's ongoing unfolding. In this way it becomes clear that drawing and painting have nothing to do with the representation, from without, of a separate reality, rather they are better understood as the world's expression of itself from within itself.[57]

Returning to my caving experience. What had ahead of time felt like quite familiar methodological territory, I had assumed we would use various recording devices to generate material, data, about a site, that would then become part of later artworks, felt in-the-doing different. Accounts of creative practices as generating rich and full accounts of a field waiting to be drawn into view, failed in the face of my experiences of these fieldworkings. To explore further, I want to turn to a second field experience – a four day trip to the Chamonix Valley, France, to look at ice-caves.

Field 2: Glacier

We stood excited, twitchy, in the entrance to the kit shop where we were meeting our guide. We collected crampons, and stomped around trying on the boots that we needed for the crampons to fix onto. We fiddled with the buckles of harnesses unfamiliar with their ways, we brandished ice axes with more aplomb than I at least felt. What we did not immediately don we attached to various loops and hooks on our rucksacks, or stowed inside next to cameras, roles of blue paper, fabric, a flask of blue liquid, and a series of shaped metal vessels.

We drove down to the station and boarded the rack and pinion railway alongside others carrying similar kit as well as families in colourful wool, various fleeces, and wellingtons or trainers, with cameras around necks or camera phones at the ready, the odd selfie stick. We discussed the plan for the day, what did we want to do, our guide wanted to know? Go onto the glacier, what would we be doing there?... umm- silence, I was not quite sure, looking around, hanging out, trying out some art? It seemed hard to explain – I did not, to be frank, really know. It had seemed quite clear, we would hire a guide, go onto the glacier – but now I was hazy, suddenly deeply unsure what I was doing.

The railway snaked up through the wood, and eventually emerged onto the site of the mountain above the bare rock and ice of the Mer de Glace, to the side of the track was the hotel where Mary Shelley had stayed to compose Frankenstein. We descended from the train, and walked towards the viewing station, around past the café, and down the tar-macked road, until it gave way to a path cut into the side of the rock. This got narrow and narrower, snaking down and down, past where the surface of the glacier would once have been, before eventually giving way to ladders, 400ft of ladders clinging to the side of the valley, down to the ice below. After a short instruction about via ferrata techniques – a series of fixed lanyards and clips on a designated climbing route – we negotiated our way down, slowly, it was daunting. The ladders were cold and a little slippery; sometimes getting on and off was tricky, necessitating an awkward stretch, a short dangle and drop, or a sideways shimmy while clinging onto the rock face. Always remembering of course to move your rope and clip and unclip yourself around the anchor points. We finally made it down to the edge of the ice. Now the difficult bit, getting onto the glacier.

Field Work: The Mer de Glace, 2017. (Image, author's own)

Once on, we could look around more, it was still hard going, uneven, slippery when rock debris gave way to ice, the crampons helped. We had seen a large erratic, poised part way up the glacier, that became our goal. Had it been the one that John Ruskin had sat on we wondered? Was it the one visible in the images of the Mer de Glace? Had Mary Shelley seen this rock? I felt a little silly in all my kit, having only the day before been looking at the classic images of Victorian women exploring the glacier full skirted in their crinolines, using their fabric umbrellas for balance. We reached the rock. Put down all our bags and dug out some nuts, raisins, ginger tea, and bizarre orange crisps. Flora began to unpack, the roll of paper, the fabric, some twine and tape, blue water and the vessels. I was not sure what to do. I watched, fabric was stretched, blue liquid was poured, paper was rolled out, taped and held down, photos were taken. Fabric was adjusted, moved into a small cavity beneath the boulder, an action that involved navigating the incredibly slippery ice.

We moved around the rock, dipped our hands in the pools of water on the top of the glacier, musing on how far down they might go, some not far, light blue at their base, others dark and deep (Figure 1.5). We turned around and walked back to the edge of the glacier, more photos were taken. We clipped back onto the ropes. The wind had picked up, rain was sheeting across the side of the rock, it was even more of an engaging climb going up.

The following day, clagged in, we continued our research in the town. We entered the museum; there was a model relief of the glacier in a box. Next to it some scientific instruments, around the corner an old board game in a case – Climb Mt Blanc – and then many photographs, sketches, and paintings. The glacier emerged from the displays as a site of

FIGURE 1.5 Field Work: The Mer de Glace, 2017. (Image, author's own)

science and art, playing a founding place in the history of 'en plein air' art as well as being key to the histories of glacial science as a field practice – an empiricism also based on being there. I knew what to do here. I took photographs, made notes, pored over the materials in the cases, listened to audio-guides, watched videos, and reflected on politics of display and narratives of presentation. I mused on the aesthetic effect of projecting rocky surfaces onto smooth floors and the balance of art, science, and exploration on display.

Account of fieldwork in Chamonix, France, with Flora Parrott, Luce Choules, and guide Angelika.[58]

If going underground felt like we were negotiating the legacies of a highly masculine relationship to the landscape, caricatured in the posturing of our guide, the glacier-based fieldwork felt equally haunted by an historical model of a different kind of fieldwork. Most obviously this felt like an expeditionary practice – after all the Mer de Glace is the largest glacier of the Western Alps. Yet as my research in the museums and afterwards made clear, the site has long been an important location in the evolution of field-based practices of science and art. The glacier was the location of many pioneering studies of eighteenth- and nineteenth-century glaciology, and was the subject of one of the most extensive sets of field-sketches of any European Glacier. These latter have offered valuable sources for those seeking to model glacier advance and retreat in the absence of scientific record.[59] Equally, as the museum displays throughout Chamonix and nearby village of Argentieres, made clear, the glacier has long drawn artists. Indeed, it and the ice cave at its mouth became a key site in the evolution of *en-plein-air* painting practices. These foregrounded on-the-spot observation of the facts of light, colour, atmospheric conditions, and landforms as a principal subject for art. Drawn in the midst of landscapes in flux, these ways of working were, at the time, quite radical challenges to more classically derived notions of form, order, and universality in art.

Some of the fieldwork we did on that trip felt familiar. We went to galleries and museums, explored archives and examined the ways the glacier and the ice cave were described and presented. We collected data in the form of notes and images on the museum displays. We conducted informal and *ad-hoc* interviews with locals – from hoteliers to photographers. There was a lot of walking, photographing, and talking in the landscapes we moved through. We shared different bodies of information; Luce, an artist we were on the trip with, knew the area well and acted as guide. Whist Flora and I added information gleaned from diverse secondary sources ahead of the visit, whether these be scientific papers about the valley glaciers or websites and YouTube links. In my case, my experience was haunted by an earlier visit to the valley and the Mer de Glace as a teenager on school geography fieldtrip. These kinds of field practices made sense to me. I could understand them as part of an existing model of what 'data collection' was. I could amass a new body of information, contemporary and historical about, for example, how an ice cave had emerged at the glacier snout during the Little Ice Age, evolved into a site for science and tourism, and then how with its demise a new practice of digging a 'fake' ice cave mid-way up the glacier had become an annual occurrence ensuring tourists would continue to visit the site.

As well as situating these familiar practices of data collection, I could also make sense of our varied activities over the four days as part of the aesthetic practice of 'deep mapping.' Deep mapping in the way I was thinking about it was as an inter-disciplinary arts practice, commonly described as a loose gathering of materials and techniques to form an intensive 'topographical exploration that aims to present diverse source – histories, economies, poetics, memories, and so on as being equally valid.'[60] In our case, to think of our visit as conducting a form of deep mapping enabled me to make sense methodologically of our forms of data collection and the resulting body of material I came away with – from old maps, and slides, to drawings, photographs, documents, sound recordings, rubbings, and collections of natural and man-made objects. The concept of 'deep mapping' as an approach to place, has been, as Selina Springett explains, deployed as both a descriptor of a specific suite of creative works – deep maps – and as a set of aesthetic practices that creates them. I would elaborate this to include a sense of deep mapping as method.[61] While its definitions have been amorphous and adaptive, one enduring feature is deep mapping's emphasis on a democratisation of knowledge. This takes place through a 'flattening' of knowledge systems, and a practice of assembly of materials (oftentimes literally through practices of collage and montage) that foregrounds productive disjunction, and celebrates difference through a 'crossing of temporal, spatial and disciplinary boundaries.'[62] As archaeologists Mike Pearson and Michael Shanks suggest in their book *Theatre/Archaeology*, this is a 're-oriented and quintessentially interdisciplinary view of landscape, one that pays heed to the grain and patina of place . . . the interpenetrations of the historical and the contemporary, the political and the poetic, the factual and the fictional, the discursive and the sensual.'[63] For feminist art theorist, artist, and long-term friend of geographers, Iain Biggs, deep mapping is an innately feminist practice evolving a form of essaying as performative resistance. This is an essaying that is not only about words, but one that that offers a 'new ecology of embodied knowing.'[64] Through Biggs's practice and writings, deep mapping emerges as a form of geopoetics that cultivates a relational way of knowing through underscoring the fundamental connectivity of various knowledge orders. To make sense of our work in Chamonix in these ways felt easy and familiar. Yet the day on the glacier was discomforting, it was not immediately clear to me how I was to understand this as fieldwork. I later learnt that Flora had also struggled to make sense of this field day too.[65]

Looking back at my notes, one way to understand the focus of the day was around conducting a series of experiments in the field with materials. The terms on which these were conducted were unclear to me. Some had their origin stories which Flora shared, a picnic held in an ice cave during an exploration she had read about, there was a glass of liquid, some paper, some delicately shaped vessels, made of metal their materiality was much more substantial than their form suggested. Fabric was placed in hollows in the ice, images were made. Bodies navigated the slippery ice, we had snacks and drinks. We rolled out blue paper. We found some abandoned small metal canisters and tins, poised for display on a rock; we sunk our hands into punishingly cold meltwater pools on the glacier surface, rapidly taking photos of our submerged digits and limbs, wriggling them in the cold air afterwards. We reflected on being on-top of the ice rather than within it, contrasting this with the tourist ice cave we had been in the previous day.[66] I felt anxious, a bit fidgety, somewhat useless. There was not really any 'data' being collected – photographs were being taken, but that was about it. My field notes from that day are a confused, halting set of reflections, shaped by failure and my worries over a lack of plan, and sense I should have been doing something, but I was not really sure what. I was trying to find within these creative fieldworkings the same set of practices and protocols of method I was trained to associate with fieldwork, and that I had been able to find in other creative field practices shaped by drawing, sound recording, photography or even painting. After all, in the field you made notes, you conducted an interview, you took pictures, you collected things, you made recordings, or even rubbings and drawings. Without these clear procedures, I felt lost, grounded only by my sense of watching the others to see what they did.

Watching Flora and Luce in the field, suggested they too lacked protocols to follow, it was not just me. They were not, it appeared, using drawing, sound-recording or photography in ways that would extract data about that place. Nor were these being used as techniques to somehow mediate their relationship with that place. Seeking ways to make sense of what I observed and felt, and what my notes utterly failed to make sense of, I came across Derek McCormack's musings on his experiences with the Montreal SenseLab.[67] Somewhat confused by how the Lab's research-creation events he attends are research, he proceeds to explore these as events as 'thinking-spaces.' Akin with the stumbling failures of my own field-notes to record anything other than confusion and my own sense of failure, McCormack notes that one of the difficulties he encountered 'might be in telling the story of what happened, a difficulty in trying to make too many things cohere and add up even when they don't.'[68] His solution is to consider the after-effects. In a sense it was the after-effects of the experience and the challenges it posed that made me want to return to the field experiences and try to fathom them. Writing of these 'thinking-spaces' McCormack posits a form of research premised not on extraction or on protocols that have a pretence towards objectivity, detachment, and disembodied distance. Instead, understanding the lab events as thinking-space enables him to appreciate research as a process in which 'the world participated in the movement of its own becoming.'[69] Field as thinking-space means thinking about the field as a site of encounter with something in the world that forces us to think, the field as a space of 'distraction, agency, action and encounter, within which research materials are less discovered than co-generated.'[70]

Framing my field experiences through this sense of thinking-space seemed to help me fathom my unmooring on the ice. The field in these terms is less a site for the enactment of a series of protocols and practices, however much one might deviate from these, or draw in and adapt creative practices as part of the research methods enacted. Instead, what seems to emerge is a creative field, in which the goal was to experience and experiment with the field space in ways led by how it make us think-feel. As such, it was therefore alright to be drawn to this pool of water, or that erratic. It was appropriate to conduct partial experiments with the odd space beneath a bolder, to muse and play and explore the field in response to it. This still feels dangerously light-hearted. As if this can't possibly be 'real' field-work. I am left with a sense that maybe the techniques of creative fieldwork are less those of the mastery of creative practices – a taking of drawing lessons, or a learning to listen better – than they might be a disposition towards the field. Creative fieldworkings here being a getting comfortable with fieldwork as a kind of going with the momentum of unrolling processes: an ongoing thinking-feeling in the field that others might call immanence. If this was it, I felt really bad at it; it made me very uncomfortable, fraudulent. I could not really conceive of how to convince my social science colleagues this was fieldwork, let alone fieldwork that would generate something called research. Nor however, could I really feel comfortable with my creative collaborators, as I was overcome by my desire to feel like I was doing something, being useful, I was unable to be present and re-sponsive to the field itself (despite years of yoga classes in which I worked, almost daily, on being present). I wondered if you could learn to be better it; could I let go of my disciplining and training to sit more easily with the field in this way? Could I ever get to a point where I would feel like in not-knowing where to di-rect my attention I was not constantly failing to attend to the right thing? Would I ever stop wanting to foreclose this thinking-feeling with a pushing or a rushing at certain kinds of known methods and approaches, or even theories, too quickly closing down any uncertainty or response to what was actually in front of me?

There were also more practical considerations at work too. What does it mean for activities to unfold in the field, rather than being directed towards fixed ends? How do you plan for this sort of fieldwork, how does it mesh with the kinds of institutional structures now required of research, from ethics forms, to health and safety plan-ning? How would I 'safeguard' in the current language of UK institutions, myself, my team, and the participants? Importantly, looking back and across my notes, this was clearly not an anything goes experience; conditions were placed by physical and human factors. What planning fieldwork 'without a plan' seemed to mean, was not quite the airy fairy, wishy-washy, go with the flow kind of sensibility and experience that the above reflections are perhaps worrying over. Rather, on reviewing the ex-perience what retrospectively emerged were a series of forms of constraint – human and non-human – that placed limits on our activities. We had decided roughly were we were going, but this was a dangerous environment, we were roped up, our route limited by physical geography and skill, the weather could change at any moment. We were drawn by the myth of Ruskin's bounder; Flora had brought her objects, which were deployed in a range of settings in relation to ice, rock, water, each other, and us. We had cameras and iPhones. Thus fieldworkings here were also perhaps a practice of context generation, of constraint setting and negotiation – by ourselves and the environment – that then enabled the experience to unfold.

Creative fieldworkings

Our encounters with the caves and the glacier suggest a variety of different forms and approaches to creative fieldworkings, rather than one singular sense of the field. Across the discussion three key ideas emerged. First, the figure of the creative fieldworker within geography as a fieldworker who uses devices – cameras, sound recorders, paint-brushes, pencils – and creative practices to mediate certain kinds of experiences and accounts of the field. Those using creative methods within the field have interestingly tied the discussion of these mediums and practices – from sound recording to writing and drawing – into the evolution of different kinds of relationships between researcher's body and world. From these relationships, spring different imaginations of the field – primarily based on proximity and openness, and even a vulnerable relationship between researcher and field. A second kind of fieldwork that emerged was the aesthetic practice of deep mapping. This enabled an understanding of the use of a myriad of techniques to get to know place and the valorisation of aesthetics as a lens through which to appreciate the value of bringing together very different registers of knowledge in a way that did not necessarily subscribe to hierarchies of thought. Yet what was clear was that my experiences during the guided caving trip and on the glacier could not be exhausted by these registers of fieldwork. Exploring the cave with these practitioners, watching how their work evolved as a result of these experiences, thinking through the kinds of 'data' that was 'collected' on the day, suggests an emerging form of artist fieldworker – other than that as truth-teller – and a somewhat different disposition towards the field. My experiences of artistic fieldwork on the glacier helped think about these ideas further.

What emerged was a need to complicate fieldwork beyond the practices of instigating a series of research methods, according to pre-planned protocols. Instead, what was important seemed to be the creation of the context and the generation of a series of productive constraints that enabled experiencing and experimenting with the location at hand. This is to think of the field, following McCormack as a 'thinking-space' where creative fieldworkings emerge as a disposition towards site that enables a responsive openness towards the field as object of encounter. As JD Dewsbury productively suggests, our field methods become 'less a question of what we do' and 'much more a question of how we are going to configure the world.'[71] As Pamela Richardson-Ngwenya observes of her experiences of the success and failure of video methods to investigate 'vitalism' what was less important was the techniques, and more important was an imagination that was open to vitalism.[72] The field emerges in this accounting of creative fieldwork, less then as an encounter in which data is extracted and taken away, and rather as an experience of collective thinking with the site, where we are alive and responsive to the world around us, experiencing and experimenting with it, imaginatively open to it.

Notes

1 See Professor Schreve's reports on https://heritagerecords.nationaltrust.org.uk/HBSMR/MonRecord.aspx?uid=MNA139958 (accessed 2/2/2020).
2 This fieldwork was conducted throughout 2016 and 2017 during Flora's Leverhulme Artist in Residence Grant.
3 Driver, *Geography Militant*; Mott and Roberts, "Not Everyone Has (the) Balls"; Staheli and Lawson, "A Discussion."
4 Driver, *Geography Militant*; Driver, "Editorial."
5 Driver, *Geography Militant*, p. 268.
6 Driver, *Geography Militant*; Mott and Roberts, "Not Everyone Has (the) Balls"; Staheli and Lawson, "A Discussion."
7 See for example, https://capefarewell.com; http://www.unknownfieldsdivision.com/mission.html (accessed 20/12/2019). Arends, *Chrystel Lebas*.
8 Lebas, *Field Studies*; Vilches, "The Art of Archaeology."
9 Cape Farewell, https://capefarewell.com (accessed 20/12/2019); see also Straughan and Dixon, "Rhythm."
10 See for example Mark Doin's 'Field Guide to Curiosity,' online catalogue at the V and A. https://www.vam.ac.uk/articles/a-field-guide-to-curiosity-a-mark-dion-project#?c=&m=&s=&cv=&xywh=-688%2C-753%2C13759%2C9479 (accessed 20/12/2019); Latour, *Reset Modernity*; Magrane and Cokinos, *The Sonoran Desert*.
11 http://www.floraparrott.com https://geohumanitiesforum.org/event-these-pits-and-abysses-exploring-the-subtteranean/;http://geohumanitiesforum.org/project-trying-to-organise-caves/ (accessed 20/12/2019).
12 Driver, *Geography Militant*. See also Buchanan et al., "Field Philosophy and Other Experiments."
13 Bracken and Mawdsley, "Muddy Glee," p. 280.
14 Driver, *Geography Militant*.
15 Haraway, "Situated Knowledges"; Livingstone, *Putting Science*, p. 19; Greenhough, "Tales of an Island Laboratory"; Hinchliffe, "Indeterminacy in-decisions."
16 Greenhough, "Tales of an Island Laboratory," p. 234.
17 Parrott and Hawkins, "Conversations in Caves."
18 Driver and Jones, *Hidden Histories*; Maddrell, *Complex Locations*.
19 Dewsbury and Naylor; "Practising Geographical Knowledge"; Latham and McCormack, "Thinking with Images"; Sharp, "Geography and Gender."
20 Ingold, "Bindings against Boundaries"; Marcus, "Contemporary Fieldwork."
21 Marcus and Myers, "The Traffic in Art and Culture"; Clifford and Marcus, "Writing Culture."
22 Marcus, "Contemporary Fieldwork."
23 Marcus and Myers, "The Traffic in Art and Culture"; Clifford and Marcus, "Writing Culture."
24 Ibid.
25 Marcus, "Contemporary Fieldwork."
26 Ibid., p. 270.
27 Marcus and Myers, "The Traffic in Art and Culture"; Marcus, "Contemporary Fieldwork."
28 Marcus, "Contemporary Fieldwork."
29 Ingold, *Being Alive*; Ingold, *Making*.
30 Ingold, *Making*; Ingold, *Redrawing Anthropology*.
31 Ingold, *Redrawing Anthropology*.
32 Ingold, "Bindings against Boundaries."
33 Ibid.
34 Ingold, *Redrawing Anthropology*, p. 2.
35 Higgin, 'What do we do."
36 Ingold, "Bindings against Boundaries," p. 87.

37 Bracken and Mawdlsey, "Muddy Glee"; Mott and Roberts, "Not Everyone Has (the) Balls"; Staheli and Lawson, "A Discussion."
38 Driver and Jones, *Hidden Histories*.
39 http://www.floraparrott.com (accessed 20/12/2019).
40 Cant, "Tug of Danger"; Aitken, "A Phenomenology of Caving"; and Pérez, "Lines Underground."
41 Longhurst et al., "Using 'the body' as an 'instrument of research.'"
42 Pink, *Doing Visual Ethnography*.
43 Stafford, *Voyage into Substance*.
44 Balm, "Expeditionary Art."
45 Greppi, "On the Spot"; Smith, *Imagining the Pacific*; Smith, *European Vision*.
46 Greppi, "On the Spot."
47 Ibid.
48 Stafford, *Voyage into Substance*; Smith, *Imagining the Pacific*; Smith, *European Vision*; Driver and Martins, *Tropical Visions*.
49 Scalway, "Patois."
50 Brice, "Situating Skill."
51 Ibid.
52 Gallagher, "Field Recording"; Gallagher and Prior, "Sonic Geographies."
53 Gallagher et al., "Listening Geographies."
54 Ibid.
55 Wylie and Webster, "Eye-Opener," p. 38.
56 McCormack, "Drawing Out the Lines of the Event."
57 Wylie and Webster, "Eye-Opener."
58 Field Trip to Chamonix Valley, October 2016.
59 Nussbaumer and Zumbuhl, "The Little Ice Age."
60 Biggs, "The Spaces of Deep Mapping"
61 Springett, "Going Deeper or Flatter."
62 Ibid.
63 Pearson and Shanks, *Theatre/Archaeology*, pp. 64–65.
64 Biggs, "The Spaces of Deep Mapping."
65 Parrott and Hawkins, "Conversations in Caves."
66 Hawkins, "Underground Imaginations."
67 McCormack, "Thinking Spaces."
68 McCormack, 'Thinking in Transition," p. 219.
69 Ibid.
70 McCormack, "Thinking Spaces," p. 5.
71 Dewsbury, "Seven Injunctions."
72 Richardson-Ngwenya, "Performing a More-than-Human Material Imagination."

2

STUDIO

Prologue: a GeoHumanities studio?

"Come and work with us in our studio," announces the website of *"The Experimental Geography studio: The home of GeoHumanities at the University of Oklahoma."* The studio, the website explains, offers a space for the 'meeting of cultural geography with the creative arts.' You can visit this studio space, and witness its evolution through still images and a short time-lapse video that documents the transformation of the office – room 543 on the fifth floor of The Energy Centre – into a studio for experimental geography. The first things to appear in the first ten seconds of the 45 second YouTube video are office tables and chairs; a sofa and a coffee table; some large iMacs, many with second screen, and plants. Pictures are tacked to the walls, a post-card rack appears. Also on the website is a plan of the space. Alongside several desks is detailed a large table, a dart board, some whiteboards, a wall for exhibition space, and speakers and a large flat screen monitor/ TV. Elsewhere on the website are pictures of the studio hosting ad-hoc events, presentations, discussions of work. We are urged to join the listserv, a virtual space to support the studio's activities: 'we chat, critique, share, review, present, watch videos, and otherwise commune so that our research collectively grows.' There are also images of student work up in the exhibition space and an online library. This is a multifunctional space where work is made; ideas are exchanged, debated and discussed; student and staff work is shown; and education happens.[1]*

Virtual visits to the Experimental Geography Studio

Studio geographies

If the field is one of the key sites of the 'becoming' of the geographer, then in creative practice the studio occupies an equally important (and controversial) place. The studio is a much mythologised site: from the interdisciplinary workshop of the Renaissance artist to the lonely garret of the starving modern artist and today's multifunctional spaces dominated by equipment for digital production, like the Oklahoma studio.[2] For many decades now, we have been warned of the 'demise' and even the extinction of the studio, with the shifting geographies of the production and consumption of site-specific, live, and performance art leading to claims of a post-studio practice.[3] Indeed, for some the rise of art and research relations in an academic context, have further challenged the place of the studio. For Wouter Davidts and Kim Paice, the shifting role of studio 'embodies the gradual blurring of the distinction between artistic and academic activities . . . that permeate contemporary artistic ways of making.'[4] Yet, the threat to the studio does not seem to have materialised. Indeed, some contend we are now in a post-post-studio age.[5] Whatever understanding of the studio is at work, the importance of the space for many art historians lies in its status as a location through which to view changing modes of artistic production. The studio is, in short, a sort of barometer for the shifting and morphing of what is understood to be 'art.' This is witnessed in discussions of everything from the Renaissance shift from the term stanza to studio to what Caroline Jones writes of as Andy Warhol's 'machine in the studio' and more recently what has been described as the apparent move 'from studio to situation' in the era of site-specific art.[6] As Marc Gotlieb succinctly puts it, there have been increased efforts 'to explore diverse symbolic, sociological, and semiotic concerns raised by studio subjects, thematizing as they do the changing terms and conditions of artistic practices.'[7] Along these lines, although resisting any sense of environmental determinism that sometimes shapes this work, this chapter explores a series of GeoHumanities studios.

Geographers have long been drawn to the studio, asserting both its value as a site at which to make geographic knowledge and as an important site in the production of creative forms: from art to fashion and music.[8] Across diverse cultural, economic, social, and historical geographies of the studio, the space has emerged as one with multiple functions (e.g. a dream space, an archive, a library, a workshop), a mobile and networked space, as well as, as a space whose existence is constitutive of a practitioner's identity.[9] Furthermore, similar to art historical accounts, the studio emerges in geographical literature as a co-constitutive site closely related to shifts in economies of production and consumption. This would include, for example, Andrew Leyshon's account of the role of bedroom and garage-based music studios, or Chris Gibson and Andrew Warren's accounts of the changing form of the studio-workshop where surfboards are sculpted.[10] In addition, a small group of what might be called 'studio ethnographies' have emerged, exploring both solo-practitioners, as well as large 'transdisciplinary' studio teams.[11] Oftentimes, these latter accounts situate the role and agency of the material and the non-human in the working of the studio – from materials to technologies and even spiders – as much as the role of humans with a range of different skill-sets and knowledge.[12]

Given the importance of the studio space to the making of work and in the constitution of something called 'art,' it should perhaps have been no surprise that studios, and studio talk – whether that be about the lack of studios, or their creation – emerged as a reoccurring theme of discussion during this research. A diversity of studio spaces emerged, some people had actual rooms designated as 'studios' in their homes, others hired small (or large) spaces or did residencies that came with studio space. Some reflected on mobile studios – which often meant principally lap-tops – others talked about the desire for a studio. Sometimes, this was about access to space to enable certain kinds of practices- such as print-making or sculpting; for others it was about something more akin to a space to think, to lay out work and have it around. Fewer were lucky enough to have access to an institutional studio space, including like the Oklahoma one that opened this chapter. This is perhaps not a surprise as even within creative disciplines the university provision for studio space is increasingly under threat.

If one characteristic of the GeoHumanities studio is its diversity of forms, another is its fluidity, and sometimes improvisational nature. I was witness to flexible creative workspaces being turned into temporary studios as we wrote grants or exhibition text; worked numerous times on creative projects on people's floors, sofas and at tables, and even in caravans and tents. Often we were surrounded as much by books as by bits of fabric, metal or wood, as well as by laptops. I have also worked on projects that included the making of a studio as part of the installation of the work itself. Oftentimes, questions of studio space came up in the early stages of interactions of creative practitioners with geography as a discipline. This included in discussions of PhD studentships which in my department don't come with access to studio space, and my university lacks an art department to even try to negotiate such access with. Another occasion in which discussion of studio space was intense was during reflections on artists-in-residence in geography departments. Oftentimes, as the chapter on residency explores further, residencies offered points of reflection on working practices and spaces in studios, offices, desks, seminar rooms, and laboratories. On one occasion, I was brought up short during a discussion with one artist-geographer who started asking me about my own practice. I demurred, saying I did not have a solo studio practice, only a collaborative one. This led to a discussion about what I meant by studio practice and how I reconciled the romantic associations of the studio with a certain kind of material making with so-called dematerialised arts practices. The latter often concerned with live art, performance work and social relations as medium and aesthetic form.[13] While I had long written about these forms of work I had not really probed why it was I still associated artistic practice with a studio space. Across all these conversations it became clear that to query the diverse forms of the GeoHumanities studio was a powerful way into the midst of questions around the nature of practice and research relations.

This chapter explores three very different, but reoccurring forms of studio I encountered across my work with those operating at the intersection of creative practices and geographical research. In doing so, it probes evolving relationships between creative practice, geographical research, and studio form. It does so through a series of studio visits, actual and virtual (through videos or other secondary sources) across London, the US, Europe, and Africa. As such it asks the following questions: what are the ecologies and topographies of the studios occupied by those working at the intersection of creative practice and research? What ideas of the production of art and the nature of artistic labour emerge from these studio spaces? And how do the practices of these studio spaces intersect with wider studio debates, including the long-held binary between studio and study, the studio as public/private space, and the studio as the space for a sole practitioner or a collective creative space?

Unsurprisingly, to arrive at a single, coherent image or understanding of the GeoHumanities studio is not this chapter's aim. This is both unrealistic, and not a useful way to appreciate the space's functional complexity or the diversity of creative production and its varied needs for space and equipment. Discussion instead proceeds through three studio forms, the first engages with the studio and study relationship and the shifting ways of understanding the division of labour between 'research' and material making practices. The second turns to explore the transdisciplinary studio, and the place of research and the researcher in large studio teams. While the third explores the emergence of the project-based studio and its association with the form of site-specific artwork. Between them these studio forms register strongly with three of the key forms of art practice and geographical research relations: one focussed on material making; one that foregrounds transdisciplinarity; and a third that orientates around events and social relations, such as community and live art practices – rather than the production of artefacts.

Studio visit 1: 'the smell of the studio' – a site for material thinking?

We sat at the laptop in the corner of the studio, under the big windows. There is an enviably expansive table, some large shelves, sort of room dividers, sets of drawers the scale of mapping chests, another table. I am utterly enchanted. It is the sort of space I imagine an artist to work in, the sort of space I want to have to work in. I was drawn to Flora's work, not just for one of its themes – the underground which we have in common – but because of its materiality. It was once written of the work of modernist sculptor Eva Hesse that 'there was the smell of the studio about it,' denoting a certain assertiveness of its made qualities.[14] There is something of this in Flora's work which bears in its forms and materials the labours, paces, and places of its making. Further, whether textiles, sculpture, print, drawing, or collage each element individually and when brought together collectively seems to create space – an enclosed canopy, a chalked outline, a fanned form – each sketching materially or in our mind's eye the dimensions of a space. The more time I spend in Flora's studio, the more I come to think of it as one of the most powerful spaces she produces (Figure 2.1).

We are writing a grant together. It is the first (of many it turns out) things we end up working on. As we work we look at books, or search online. We check references and copy quotes from print-outs. As we explore ideas, and especially

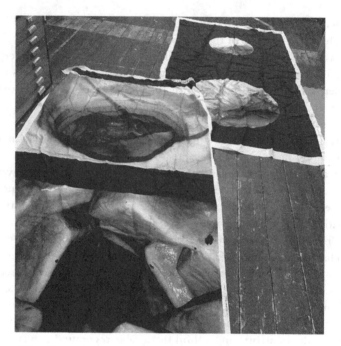

FIGURE 2.1 Flora Parrott's studio. (Image, author's own)

in the beginning before I know her work well, Flora would go to cabinets and extract drawings, or bring me to objects to hold. She would talk me through hanging pieces of fabric, or large paper prints. We would unfold textiles and paper, suspend things, or stand on tip-toe, holding fabric at arms-length, letting it waterfall to the floor.

Later, during Flora's PhD, co-supervised by myself and a print-maker at London's Royal College of Art, there is talk of access to workshops and studio-based supervision. Normally we meet in the tarted up class-rooms of Royal Holloway Geography's London base – newly re-painted in refined shades of grey, with contemporary chandlers on dimmer switches. The practice so integral to the thesis is brought into this space on various small screens, we peer at laptops and dab at iPads – our discussions feel restricted, more easily turning to text, to issues of research questions. We talk about the importance of the studio, the demise of studio space for many London art schools; Flora remarks that sometimes people end up producing work that looks like it was made on their kitchen table. I can imagine how this could happen.

The 'smell of the studio' is an evocative phrase and brings to mind the studio as a material workshop, a site of skilled material practice where artefacts are made.[15] Yet in Flora's case, in the same room where much of this was going on, were books, laptops, journal articles, highlighted papers, post-it notes, and so on. Trying to make sense of these working practices I got stuck. Over and over again I came up against the entrenched binary pair of study and study, the former assumed as a site for fabrication, and the latter for conceptualisation.[16] Baur, for example, uses Renaissance room inventories to map together lexical and functional shifts from the bottega (usually a commercial space open onto the street), to the stanza (meaning just a room) to finally the studio.[17] This shift has been related to the artist's rising social aspirations as they sought to distinguish themselves from artisans. The understanding of 'art,' moving from more economic practices such as lace-making or carving, towards those practices based on intellectual discussions, social assemblies, the discussion of ideas, or drawing from life.[18] This was an era in which skills of hand and mind increasingly came to be appraised differently, with the artists skills of scholarly, conceptual, and intellectual rigour, coming to be valued over the artisan's handed skills with tools and knowledge of materials.[19] The modern studio thus emerged as a space for the artist-scholar, foregrounding intellect rather than craft.[20] This spatial division of labour continued down the centuries, coming to a head in Lucy Lippard's and John Chandler's essay on conceptual art of the mid-twentieth century, *The Dematerialisation of Art*. This essay famously understood the evolution of the conceptual art of the 1960s in terms of a geographic shift from studio to study.[21] They write,

> the anti-intellectual, emotion/intuitive processes of art-making characteristic of the last two decades have begun to give way to an ultra-conceptual art that emphasises the thinking process almost exclusively. As more and more work is designed in the studio but executed elsewhere by professional craftsmen, as the object becomes merely the end product, a number of artists are losing interest in the physical evolution of the work of art. The studio is again becoming a study.[22]

Interestingly, and in line with my experiences with Flora and others, more recent scholarship details a more complex set of spatial understandings than the association of studio and study as respectively the spatial expression of making and thinking. Performative reconstructions and material culture analysis of everything from medieval workshops to Renaissance rooms foreground 'lost' studio models that re-narrate any separation of skilled material practice and intellectual labour.[23] Michael Cole and Mary Pardo, for example, describe the studio as originally an intellectual site, whose properties infiltrated and transformed the artist's working space.[24] While, in the Sixteenth and Seventeenth centuries, the studio served as the location for what was then thought of as the creative part of an artist's activity (as set apart from the final material practice). That is, 'invention through drawing and modelling as distinct from execution in paint or stone.'[25] During this era, drawing and design were understood as 'visual thinking,' manifestations of an artist's intellectual labour alongside the investigation of ideas and the compilation of thoughts.[26] Renaissance historians have explored the combined evolution of the sketch and the studio-as-study, arguing that the sketch emerged as a new form of drawing, its rapid lines capturing information that could then be copied in the workshop.[27] As such, drawing becomes situated as a meditative form between the intellectual thought processes, and the material renditions of that process. To excavate these histories of intersecting spaces of making and thinking is to encounter the complexity of a historically situated sense of what counted as 'art.' In doing so, it directs our attention to the evolving intersections of, and distinctions between intellectual labour and artisanal labour as part of these discussions. Yet reflecting on the intersections of material making practices (sewing, printing, painting, soldering) and the obvious manifestations of 'intellectual' practices, reading books, working on a laptop, writing by hand or on-screen the distinctions felt a lot less clear. To elaborate further, I want to push at the binary of thinking and making that these re-tellings of studio history enable. I want to explore how, in working with Flora and others, it seemed clear that the sculptural forms, the experimental fabric elements, the folded pieces of paper, are very much a part of how the research process proceeds. In other words, creative practices are not only method to gather data, they are not primarily something that comes after 'thinking,' not simply the execution of a conceptual idea in material form, but are the thinking itself. To consider this further I want to draw on the concept of 'studio-works' that art historian Bryony Fer evolves through her study of conceptual sculptor Eva Hesse.[28]

Eva Hesse (1936–1970) was a successful multi-media artist, who shortly before her untimely death from a brain tumour in her mid-thirties turned to sculpture.[29] Her experiments with material and form have shaped her place in the largely male-dominated sculptural canon of that era. Whilst many of her colleagues and peers were working in the land or with huge durable forms, she turned to small-scale works and more ephemeral materials – ropes, cotton, gauze, rubber, dyes, and so on.[30] Over time these broke down, decayed, and rotted, changing colour and form: posing significant issues about how to study works whose changing form was integral to their conception.[31] A key film made about Hesse within her studio demonstrates the very flexible ways in which she imagined her works could be displayed. These often foregrounded the 'in process' nature of her practice and the studio as a site of display, worlds away from the modernist white cube gallery, with its clear vitrines and plinths. Even more unconventional than these 'finished' pieces, were the so-called 'studio-works.' These were made in a range of materials, including metal, cheesecloth, latex, and twisted wire sheeting and often look like they could easily be studio-waste. These works however, defy any dismissal of them as test pieces or left-overs, instead they have, to quote Fer, the 'breathe of art' about them. Whilst other art historians had overlooked these pieces, Fer seized on their ambiguity as an important site through which to think through Hesse's practices. She queried how these 'studio-works' insist we question what we are looking at, artworks, test-pieces, maquettes?[32] Going back and forth with the status of these works, and stuck for the correct term to capture their importance, and the role she wishes them to have within and for 'art,' Fer coins the neologism 'studio works.' In doing so, her probing of the edges of these 'sub-objects,' which she describes as the thinnest of art, by the 'narrowest of margins' becomes a way to foreground the centrality of process and ongoing experimentation to Hesse's work, and the challenge this posed to the rendering of any work 'finished.'[33]

What do these studio-works as sub-objects offer me as I try to think about creative practice and research relations in the studio? Whilst many commentators on Hesse's work, in common with Fer's discussions, have drawn attention to its effects on spaces of archive and display, I want to suggest that her work can also shape how we think about the studio. First, in their uncertain status, their constant back and forth between being art or not-art, or in their being art by the thinnest of margins, these studio-works affirm the need to value creative practice as a process of material thinking and investigation. These objects like Hesse's other works are never really finished, complete, but rather are always in process, on the way towards being something else. To take this seriously is to help us appreciate the value of guarding against understanding 'art' in relation to research as only about a finished thing, only about an object presented in a gallery space for public consumption. Instead, we need to appreciate how practice-based research requires that we situate artistic practices throughout the whole research process; indeed, they are the research process. In other words, we need to move away from only appreciating art as an 'output' of the research, or as a method for data collection. Understanding each piece as part of ongoing experimentation, and an evolving body of work, of ideas, was something that working with Flora helped me accept. I had been used to thinking of journal articles and presentations as some kind of 'finished thing,' something that would of course be built upon but something that was finished in a certain way. Whereas, the events, workshops, articles, fieldtrips that we went on felt more like ongoing experiments in a wider body of work, this seemed to enable more risk-taking, more out of the box thinking.

Second, the idea of 'studio-works' not only foregrounds the sense of process, but also enables a certain kind of detailing of that process. 'Studio-works' very clearly frame, dramatise, and value the studio as a site for material thinking – in other words, Hesse's studio-works were rarely the material manifestation of pre-existing sketched forms. As such, I am helped to understand my experiences of working with Flora. Materials and objects were part of our processes of idea development, they were not supplemental to conceptualisation, rendered in their material form after the idea had been thought through, but rather were part of how ideas were worked out. Importantly though, like Hesse's works, these were not something we might call 'test pieces,' with its industrial feel of somehow creating systematic iterations until the right answer is found. Nor does the more artistic idea of the 'maquette' work. This latter option suggests a smaller study – a model – for a larger work that might follow – this is not the case. Instead, these studio-works emphasise the labour of the studio as that of a making-thinking that is an experimental practice in its own right, and that should be understood as part of the process of practice-based research. Sometimes in Flora's case these individual objects, the sculptural forms, the fabric that we worked with in the studio might reappear in other contexts within and without the studio. They might emerge as components in installations such as those described in the opening section of the chapter on field, or might be used in the field itself, as described in our work on the glacier.

Thinking through my encounters with objects in Flora's studio using Fer's idea of 'studio-works,' voids any distinction of material making practices and intellectual-conceptual practice, such as that found in histories of the studio and the artisans' workshop, or in the geographical rendering of conceptual art as shift from studio to study. Instead we are required to really appreciate process in multiple ways. Thinking with Fer's studio-works encourages us to resist any sense of closure to these projects. Instead research is a process that folds over on its self, never really ending. To think with 'studio-works' is to appreciate the myriad forms of research in the context of the studio, that material making is not necessarily the rendering of a mental image that is already present, but it is in itself an exploratory making practice. These issues of the division of labour continue to be relevant as we move into the next discussion of the transdisciplinary studio.

Studio visit 2: the transdisciplinary studio on display

The curation of Olafur Eliasson's 2019 retrospective *In Real Life* at Tate Modern, London, put the studio on display. The first room I entered was a room within a room, at the centre of which was a huge vitrine full of working models – geometric forms in metal, woven paper, and basketry, some fascinating maquettes for larger pieces that I later encountered, with small cut-out people.[34] Even to my untrained eye, the artistic and architectural references where clear; here was a series of experiments with Buckminster-Fuller's geodesic forms and its relations with atomic structures; there sat something more akin to Archigram's college form. Presented higgledy-piggledy, some upside down or lying at an angle, architectural model butting up to 3D mathematical form, roughly glued wood next to slickly printed plastic. This presentation did not really invite me to examine each of these as individual objects, to appreciate the process step-by-step, as you sometimes see in the presentation of sketch books or working models in gallery spaces. Instead, the room-within-a-room felt more like presentation of process as spectacle. There was no explanatory text (Figure 2.2).

At the opposite end of the exhibition, entering the final room, standing beneath a big exit sign with an arrow I was faced with a caption – 'The Expanded Studio.' The Expanded Studio, we are told,

> evokes the broader interests and activities that Eliasson's studio in Berlin pursues. The long pin board wall here is based on the walls in the studio where teams of researchers and craftspeople, as well as Eliasson himself, share questions, articles, images and news clippings... Every other Wednesday there will be a live link up with the studio showing daily life and a range of projects and activities taking place there.

FIGURE 2.2 Installation view of *In Real Life*. (Image, author's own)

Following the very neat recreation of that pin-board wall from right to left is to
go on an intellectual journey from Atmosphere to Zoom, by way of Data, Experience, Uncertainty, and Volcano. Material is organised as a set of alphabetised
studio 'key-words.' Glossy printed photocopies of book covers, of the front pages
and abstracts of journal articles, sketches and images, complete with annotations
are neatly tacked to the wall. Accounts of fieldtrips, pictures of people in the
landscape, the lab, and the studio join the text. There are no ripped edges here,
only artful highlighting and neat notes: a somewhat aesthetically pleasing and hy-
gienically presented set of inspirations. Hand-annotations detail the full sources
of PhDs or useful articles. At the end of the wall, the college petering out, a
hand-written yellow sign informs us with a studied nonchalance that we the full
list of references could be found online at https://olafureliasson.net/inreallifere-
sources/. This research aesthetics/aestheticised research process might have been
presented as the thought processes of the studio, but it could equally have been
a snapshot of the bookshelves of many a cultural geographer. The inspirations
read like a who's who of Anthropocene thinking: Elizabeth Povinelli and Donna
Haraway; the Whole Earth Catalogue and its various contemporary remixes;
Henri Bergson on Creativity; phenomenological and post-phenomenological
ponderings on experience and perception; and a smattering of Sloterdijik, with
Extinction Rebellion thrown in for good measure.

In a sort of passage between the 'Expanded Studio' and the exhibition shop,
a shelf of books bears the caption 'The Expanded Studio Book selection.' Here

original philosophical texts sit alongside monographs on Eliasson, cook-books about slow cooking, the latest popular climate change polemics, various Jane Bennett and Donna Haraway texts, and Anna Tsing's 'Mushroom Book.' The caption details:

> Olafaur Eliasson's practice extends beyond making artworks, installations and public sculpture to include projects that address questions facing the world today to do with climate change, migration and renewable energy. In order to capture this dimension of Eliasson's work and to elaborate on his interest in intellectual exchange and dialogue Tate Bookshop has collaborated with Eliasson's studio to create this unique and of-the-moment selection of wider reading around the ideas and subjects explored in The Expanded Studio.[35]

Eliasson and his curators had chosen to book-end this survey of his work by a display of the processes of Studio Olafauar Eliasson (hereafter as it is known SOE). It is no exaggeration to suggest that his studio is one of the most successful and enduring of Eliasson's works; its forms and working practices offering as much a site for critique as its various outputs – from huge installations to almost 1 million 'Little Sun' solar energy 'sculptures.'[36] As accounts emphasise, SOE is a hundred-person strong studio based in Berlin founded in 1995 that encompasses multiple forms of labour. Seemingly echoed in the encounters with SOE that the exhibition created, the studio consists of those who work in the workshops and metal-shop, welding and fabricating objects, trialling designs, making maquettes, like those seen in the 'model room.' There are also those who do 'research,' working in the 'The Research and Communications Department' whose labour finds form in that pin-board wall of the 'Expanded Studio.' This is a visualisation of what Eliasson has described as the 'trading zones amongst art, architecture, mathematics, engineering and physics' that he sees his studio as exploring.

The presentation of aestheticised literature reviews and artistic inspiration boards in gallery spaces has come in for some critique of late. In some cases, as with Eliasson's exhibition, they appear as selections of various ordered materials, with annotations, but little material to really engage with. Other times, they function more like libraries with entire copies of books and journal articles to read and borrow, and even to photocopy and take away.[37] We could interpret these displays as attempts to share the rich conceptual sources of the work, or to ensure that the theoretical labour of art making is given space in galleries alongside the more traditional presentation of sketchbooks and models. Others, however, have interpreted these forms of display, as about a kind of credentialising of art in terms of contemporary social and critical theory and philosophy.[38] Here I am interested in what these and other displays of the studio suggest for our understanding of Eliasson's transdisciplinary studio and the forms of research that it incorporates.

"@StudioOlafuarEliasson" (founded in 1995) rewards time spent in exploration despite, as co-editor of a recent studio-expose book 'Open House' suggests, raised eyebrows in the art world regarding Eliasson's practice.[39] Indeed, as Jones observes, 'given the compelling nature of this studio model, the world of Studio Eliasson needs to be further parsed.'[40] For not only has SOE become *the* defining and influential form of the transdisciplinary studio, a form of studio that must surely be of interest to those of us thinking about the intersection of geography and creative practice, but it has already drawn the attention of geographers. SOE, alongside another large transdisciplinary studio, Studio Saraceno, and organisations such as SYMBIOTICA and Cape Farewell have proven of interest for geographers not only for the ideas of environment, atmosphere, and cosmos they create and circulate through their works but also for the experimental and often collaborative practices that drive them.[41]

Art Historian Alex Coles included SOE as one of several examples that enabled him to elaborate on the emergence of the transdisciplinary studio – a key example of post-post-studio practice.[42] Such a studio, he writes, is characterised by a situation in which 'artists and designers are defined not by their discipline but by the fluidity with which their practices move between the fields of architecture, art and design.'[43] This is perhaps a rather limited understanding of transdisciplinarity in that only three fields, all of which could be said to closely have a practice at their heart, are enumerated here. More salient perhaps is how the different knowledge producing practices associated with transdisciplinarity come together in the studio space. Here, I move across a series of artworks SOE has produced about itself, as well as secondary data in the form of interviews and accounts of working practice, to probe further the centrality and currency accorded to the idea of 'research' as part of SOE labour practices.

Tate Modern's curation of SOE through the display of its processes was not a one-off occasion. Instead the exhibition's 'welcome' to the studio space and the invitation to interact with it, through a live feed and Twitter, felt like the spatial realisation of something long present in how Eliasson frames SOE in the context of his practice. In 2017 SOE self-published one of its ongoing series of 'cookbooks' entitled 'Open House.' As Eliasson explains in the 'welcome,'

> the book invites you into my studio in Berlin. For this book-as-openhouse my studio team and I have left some of our favourite items lying about- experimental setups, sketches, models, some artworks. You will find fragments of conversations, stray quotes for inspiration, and ideas floating about, which give a glimpse into the daily concerns and exchanges among team members and with visitors.[44]

Later he goes on to describe how he wants this to

> examine and hopefully, make intelligible the artistic processes that take place in my studio and its exchanges with the world. The book focuses on

making these processes felt from within the everyday studio work, rather than on presenting finished artworks.[45]

Caroline Jones, long-time art historian of studios, presents a persuasive argument that for Eliasson and many of his generation, the objects produced are really only part of the story of their oeuvre.[46] The story, she implies, is one of ongoing knowledge productions, which takes many forms within and beyond the studio but within which individuals, artworks, and even themed projects are just nodes. Indeed, reading across accounts of SOE it is clear that its activities take in multiple forms of production: collaborating, prototyping, reading, talking, fabricating, experimenting, outsourcing, representing, exhibiting, publishing. If part of the 'authenticity' and credentialising of Eliasson's work relates, Jones suggests, to his 'insistence on the studio,' it also perhaps rests on the ongoing assertion of 'research' as a crucial part of this studio work. Throughout *OpenHouse*, images of fabrication are interspersed not so much with 'stray quotations' as Eliasson suggests, as with entire sequences of photocopied and annotated pages of texts by Merleau-Ponty, Henri Bergson, and Bruno Latour, amongst others. The image of 'art' this creates (perhaps ironically given the force of the theory towards embodied practice and perception) is of an artistic practice grounded in intellectual currents of thought, as well as material practice. Later texts asset the political and radical purpose of art.

The centrality of 'research' as part of SOE's processes is further reinforced by the division of labour and the naming of staff that the published lists of studio personnel on the website suggests. The Summer 2018 list of studio staff noted a 'Head of Research and Communication,' with a dedicated team of 19 people, then 25% of the whole studio. Other roles included metal workers, secretaries, model makers, and cooks. The parallels between this studio-team set-up and that of the Baroque Studio has not been missed by critics, with one observing, that the 'Baroque artist was only as good as the team he had assembled, but effective in the given economy only if an "Andrea Pozzo" (for example) could still be the contractual result.'[47] Interestingly, neither Eliasson nor his team attempted seriously to present anything different. Across the volume of studio interviews and *Open House*, everyone clearly articulates how, at the end of the day, it is Eliasson they are all there for. They can shape ideas and input into work, but it is really all about what he needs and wants.[48] Eliasson appears quite conservative in his discussion of his team's work and the spatial division of this labour. He draws a distinction between the workshops and the 'office,' and in doing so reprises the division between thinking and making. Oddly perhaps, he observes, 'the translation from thinking into doing is the radical thing.'[49] In his discussion of the studio he enacts a rather telling division. He observes 'the office' as distinct from the workshop, 'the office itself is divided into rather diverse areas: I have a publication department and an archive with two or three art historians. And then there's the book-keeper and the project manager.' This is a telling grouping of functions, service-based functions – such as book –keeping and project management sit alongside research work – the publication department and the archive. This suggests a role for these art historians – the research and communication team – less as aiding in the evolution of the work itself, than in mediating it for the world. The task of these researchers is to control the framing and messaging of the work, perhaps through the selection of appropriate pages of texts to include in books and on walls.

In her long-form article on the studio for *Art Forum* Caroline Jones introduced the term 'discourse workers' to describe the actions of SOE's research and communication team. Writing of the studio as a whole, she observes 'of this group (its ranks increased by temporary workers as needed) roughly a third are skilled fabricators (carpenters, welders, electricians) a third are architects and designers and a third are discourse workers (trained art historians, an archivist, a secretary).'[50] Elaborating on this 'discourse work' she details how tasks based around the framing of the art for consumption variously as art, as science, as technology through 'wall labels, in web-chat, in exhibition catalogues, press coverage and word of mouth.'[51] The term 'discourse worker' has stuck, and continues to be used by many of those writing on Eliasson's and other similar studios. Yet looking closely, it seems that this discourse work is also being carried out by a whole host of other associated researchers, including geographers. Joining the collectors, journalists, and curators, who one might normally expect to pass through a space like SOE, are some of the great and good of European cultural theory, including Paul Virilio, Bruno Latour and his collaborator Peter Weibel, Rem Koolhaus, Mieke Bal, Johnathan Crary, Peter Sloterdijk. These individuals are often linked into SOE through exhibitions or events. Some are invited to join the experimental process or attend meetings in the studio, others write about the work, either in catalogues or in their own publications. Furthermore, looking closely at the interviews with these discourse workers, or 'in house' critics, as Coles prefers to describe them, suggests that their role is not solely about the public framing of works, or even their preservation in the archive. Rather, it seems their research function is distributed throughout SOE's production process. As one individual observes,

> I see no radical difference between the work I do and that of the architects, craftspeople or technicians in the workshop downstairs… for Olafur, I guess it's important to explore various fields of research in order to extract knowledge or a certain vocabulary that he can bring back to art so to speak. What we do is essentially art and what we are borrowing from other fields- words, molecules, whatever- we insert into at practice, transforming them and putting them to use in a way that expands what we are producing in the field of art. I continually ask myself what can I contribute to a given situation - to add to the complexity of the questions being asked.[52]

This suggests research is understood here as a kind of extractive capacity – a set of practices of identification, location, and extraction of ideas and their importation into the artistic field. When pressed details of these 'research' processes emerge as 'tracking down examples,' making 'contacts with other individuals,' and 'leafing through good books.' These source materials then come to inform the evolution of the artwork. As Eliasson describes, these ideas are always 'held against an art work, tested,' and eventually transformed. The relationship between these different forms of knowledge is presented as part of the rich maelstrom of 'team Eliasson.' Elsewhere the studio's organisational form is somewhat romantically described as 'flat,' 'anti-hierarchical in its distribution of labour, and permeable to restless streams of intelligences from all its players.' Together these are presented as contributing to the studio as being conceived of as a space for the 'resetting of problems, designs and the testing of ideas' helping them reach 'clarity and conceptual precision.'[53]

If there are multiple forms of research at work in the studio, it is interesting to reflect on the labour of the academic as part of this studio collective, whether it be those established figures coming to work with the artists, or whether it be emerging academics who have studied these studios often as part of ethnographic doctoral work. As such, figures such as Thomas Jellis, writing on SOE for instance, or Sasha Engelmann writing on Studio Saraceno come to occupy interesting roles in relation to the studio practices.[54] In both cases, studio ethnographies began as part of doctoral work, and in Sasha's case evolved into an extended relationship that has seen the production of an academic monograph about the studio, as well as her taking a role in curating and disseminating the studio's works through social media and in major exhibitions, including a series of events at the Palais de Tokyo in Paris.[55] She has also co-written book chapters and research papers with studio representatives, including Tomas himself, developed education programmes teaching with Tomas and other studio members, and has been part of the evolving open source community the studio is catalysing.[56] Engelmann's role enables an elaboration of the idea of the discourse worker, as a seeming point of collision of the academic as mediator of the artist's work, shaping its place and uptake, but also as an academic whose ideas and concepts are also clearly found in the evolution of the work itself. This is evidenced in the exchange between academic concepts and artwork that co-delivered talks by Engelmann and Saraceno demonstrate, and that emerges even more clearly than their collective writings. Interestingly though, Engelmann stands back, in her own writings to-date, from the perhaps tricky question of how to situate her own role in these complex teams, preferring instead to offer reflections on the nature of collective studio labour. Whilst she never asks the question 'what is research,' of Studio Saraceno's collective labour, she does write of one of the studio's ongoing projects as 'an extended research project in scanning and analysing the web of the black widow.' Tomas Saraceno's research team is described as involving 'arachnologists, astro-physicists, as well as the spiders themselves to help understand the homology between spider webs and the cosmic web.'[57] In Engelmann's account, the research expertise of various scientists is folded with construction expertise of the studio team, and the web-making practices of the spiders themselves. All parties are understood as contributing to the collective achievement of the artwork. Here, by way of Anne Saughvarin's ideas of 'art-machines,' itself a legacy of Deluzo-Guttarian thinking, is a conceptualisation of studio practice that situates the artist as part of a 'web' of other practitioners who are not just human, 'of spiders, simulations, scientific theories and technological devices, an entire assemblage reproducing images that exceed the work of a singular human figure.'[58]

As emphasised in the opening of this chapter, the aim is not to evolve a singular understand of the workings of geohumanities studios, nor to evolve a singular understanding of research that might be practiced in these studios. What emerges from across these two transdisciplinary studios is a clear sense that whatever different forms of practice are involved – art, architecture, engineering, as well as non-human practices – research is a term very much valued. Sometimes that research is given form and face in the figure of often high-profile academics. At other points, 'research' seems to be set up as a code for theory, or other kinds of often written-based work, that offers material and framing for the art. Elsewhere, research is the means by which the artwork proceeds. The challenge is perhaps less to define one of these forms as more valuable than another than it is to try to hold all of these forms of 'research' within the transdisciplinary studio together. What emerges quite clearly however is that for Olafaur Eliasson and perhaps as much for Tomas Saraceno (although this remains to be seen), the spatial forms and organisational labour of their studios are perhaps one of their most enduring contributions to the art world. In Eliasson's case the studio is as much a focal point of critique and academic discussion as his artworks are. In the final section, I turn to the studio of a geohumanities artist whose practice is socially engaged performance work, producing site-specific live artworks through social relations, rather than producing art objects.

The project-based studio – from studio to situation

In a welcome exit from heat of the Tunis street we pass through a heavy wooden door, set, as with many of the buildings in Tunis's old medina into a much larger door, and onto a cool tiled floor. We climb the stairs; the walls are papered with posters for arts and music events. The stone of the stairs is dark, and smoothed by many pairs of feet over the years. At the top, a room opens onto two large spaces with desks and book shelves. There is a small sitting room, kitchen and bathroom. Another large front room overlooks the street; the windows are shuttered against the heat. We are invited to sit anywhere, suggesting that the front room is the best for work, and are given the all-important internet code. We set up shop. Bags open, laptops and plugs are extracted, speakers, a dictaphone and the odd notepad are laid out across the table. Someone fetches coffee from the café down the street, we send Dali, as a man it is more straightforward for him.

Tania starts to listen to recordings of interviews, I open the file I am working on, an English translation of one of the texts for the performance. Dali brings coffee, he moves in and out of the room making phone calls. One of the DJs arrives, looking upset, they apologise for being late, they were — they explained — being hassled on the street for not looking enough like a man or a woman. We down tools, more coffee is fetched, discussion ranges through the group about the horror of this sadly everyday violence, about the challenge faced in Tunis more generally. The experience is directly related to the content of the piece, it reinforces Tania' s decision to include not just women's stories but also LGBTQAI+ stories too. Discussion flows into the ongoing debate about how the piece will work, how the stories will be presented through the choreographing of the space, how to manage the fear of performance that some of the participants are wrestling with. We listen to some of the interviews; Tania translates fragments of the Arabic for me. We reflect on the challenges of anonymity, on how the choreography might work, on the challenges for the participants of also becoming the actors, telling these very emotional stories. We discuss how some of the elements of the translations should work. I begin to transcribe a predominately English interview. The urban studies researcher from the local university arrives, we talk about the form of the old city, about the history of it as a refuge, about gender and colonialism. Dali makes some more phone calls about materials, disappears to go and source them, he phones about dinner, what do we want, where? More fish? Tania makes a call about the printing of the text, checking in with the funders of the site-specific festival the work is embedded within. We play with the mock-up of the text sheet, unfolding and resolving it, thinking about the phasing of the text, and the images, what should go where. Is the colour right? Does the framing of the text work? Are these the right images? Who will proof the French and Arabic? We leave to go and meet the women who are part of the live artwork that Tania is making in the building just down the street.

Studio work during fieldwork, Tunis, 2017[59]

Tania El Khoury is a successful live artist; she has a significant international profile, growing after her being awarded the world's largest live art prize in 2017.[60] Earlier that year she had also finished her doctoral thesis, an interdisciplinary study between drama and geography.[61] Her work, which explores themes of migration, conflict, witnessing, and trauma, especially in the Arab world, has been featured in festivals globally and she is regularly commissioned to show existing work as well as make new pieces.[62] The piece that I was temporarily part of the team for was a site-specific commission for the festival *Dream City* held every year in Tunis's old city.[63] *Dream City* specifically focusses on local issues, asking artists to make site-specific work for local presentation/display. The artists recruited are mainly, but not always, 'Arab' artists, a problematic grouping which many, like Tania, often try to resist, but which tends to mean artists with roots in the Arab world.

The *Dream City* commission, eventually titled *Unmarry Us*, is one of an increasing number of site-specific projects that Tania is asked to do.[64] In her work site is medium and topic, and geographical knowledge is central to how the works unfold. The focus of these works is often intensely local and personal stories with global resonance, stories of power and abuse that often mesh together the interpersonal and the geopolitical. The Tunis commission evolved over the course of a year, with six months of intensive research and visits. During this time — and

I have seen this happen in London, in New York and in Tunis – Tania's studio becomes what and importantly, who, is around her. She is adept at building local connections and temporary communities, finding and engaging the spaces and resources she needs. In this case we took up residence in the base of a creative collective in Tunis who let us use their flexible work space while we were there.

Tania's mobile studio practice is very similar, in a sense, to the imagination of the studio that emerges from recent Marxist art histories that have sought to determine the impact of the rise of artistic immaterial labour on historically very materially based studio practices. Pascale Gielen, for example, explores how Neoliberal and Post-Fordist conditions have affected understandings of artistic production – foregrounding individual human capital and performances, rather than the production of objects.[65] As he and others explore, the artist in the informational society is a figure whose contexts and terms of production have shifted. The studio can no longer function as a fixed, autonomous, transcendent space, but rather needs to find form as a dispersed, fluid, mobile, and interconnected one – temporary, ephemeral spaces; sites of fluid interchange, between objects, activities, and people.[66] In other words the system that was once constituted in terms of a series of successive spaces through which an art object travelled – studio, gallery, museum – now needs to be comprehended less in terms of objects than of the production, distribution, and reception of the artists themselves.[67] The form of the studio and the figure of the globally mobile networked artist are therefore considered thoroughly co-constitutive, driven by what has been termed the general 'communicational demand' of the arts. This is a form in which itineracy and circulation are privileged over fixity (which of course has its own issues in terms of who has the privilege and resources to participate in that circulation). To echo Jones's observation of Eliasson, artworks are nodes in networks of production that orbit around the artist; achievement here is less the production of discreet objects than the realisation of the networked artist.[68] In this context the studio is no longer a space of retreat, to which materials gleaned from the world are taken to produce art, rather it is a site of social relations, which the artist creates anew in each location, ensuring they are (somewhat exhaustingly), 'always plugged in and online, always accessible to and by an ever more integrated and dispersed art world.'[69]

Yet, as much as my experiences resonated with these discussions, one of the things that I find interesting about how Tania's studio is 'mobile,' 'networked,' and 'fluid' is that this is as much about the challenges she faces as a moving, mobile artist as it is the means through which she produces her intensely local work. This studio is perhaps less about material resources that can move – a laptop, an artist – and is instead the means through which an artist can build the in-depth and careful relationships with the local communities she works with. Tania's practice is, as I have explored elsewhere and will consider again in Chapter 4 on community, akin to a participatory research practice.[70] Oftentimes, as with this *Dream City* commission, Tania's site-specific studio involves the recruitment of local artists and creative practitioners as part of her projects. This establishes not only what seems to be an important ethical premise of her work, that it is informed by local people and contexts, but also ensures that she supports artists local to where she works. In the case of *Dream City*, there was at least one local researcher – in this case an urban studies scholar – and Dali, a local artist who was employed as an assistant during the festival. Dali was invaluable for project logistics, whether it be local people we needed to know, or sourcing various resources – from fabrics to food and henna, as well as the dinners his housemates made for us every night, eating and drinking on the balcony of their top-floor flat in the middle of the old town. There were also the local participants, the DJs, and a drag-queen, amongst a host of others who helped make the work happened and who featured within it.

Site-specific art, in other words, art where the location of the work is its medium, has been framed in terms of a mobility of production – 'from studio to situation.'[71] However catchy this spatial transition might be, it perhaps misses how these works, as *UnMarry Us* demonstrates, do not so much shift away from the studio, as reconstitute once more what a studio is and does. Interestingly, the idea of movement from studio to situation also risks reproducing elements of the tension geographers have long negotiated between the study and the field. In other words, where the arm chair geographer is looked down on as not doing 'proper' research, with 'the field' seen as a site of authentic research.[72] Yet as with those recent accounts which have sought to intertwine field and study, so we need not only to intertwine site and studio, but also to recognise how work that foregrounds the former, require a rethinking of the latter. Indeed, the medium of site-specific art has long been framed by discussion about appropriate ways to engage with site. Oftentimes, in an echo of anthropology's discussions of eth-nography, the longer the artistic engagement with site, the more 'in-depth,' and akin to dwelling (as long as one does not go native), the better.[73] Much critique is directed at so-called 'plop art,' where mobile artists simply reproduce their methods in one-place-after-another as they circle the world on the festival and biennale circuit. As Foster points out, however, there are also challenges with the more durational engagement with site-specific artists at times reproducing the colonial power dynamics, extractive practices, and issues of othering associated with modern ethnographic methods.[74]

Tania's studio might be one way she negotiates this geographical trap. For, in-stead of working from studio to situation, her assembly of a local studio becomes, in effect, one way she engages with each situation. This situated assembly of a studio, constituted through people rather than mainly, or only, through materi-als, seems integral to how Tania negotiates the power relations that often chal-lenge site-specific work, in particular those that concern its relations to locality.[75] During *Dream City*, Tania and her peers spent time reflecting on and discussing the challenges of festival circuits and site-specific practices, especially where the Western art world circulates through non-Western world spaces. They explored how to resist negative tendencies, and ensure when needed these challenges are surfaced within the work, rather than ignored. What emerges from Tania's work is a sense of the studio of the global artist as less a singular bounded space from which an art object gets pushed into the world, and rather a space constituted for each project through a series of often quite intensely local connections and relations. It is through these connections between humans, and between humans and non-humans, that the artist works with local researchers, lay and expert knowledge holders, and local gatekeepers to make these site-specific works happen.

GeoHumanities studios?

The GeoHumanities studio, unsurprisingly, takes many forms. These include, the individual artist-makers drawing together a series of making-thinking processes, the transdisciplinary studio and the mobile project-based artist assembling studios in a succession of global locations. Amidst all of this, the studio retains its power as a site for making, and as a site in which research in diverse forms takes place. In exploring the dynamics of these studio spaces, we get a sense of the diverse and shifting modes of production that we find within the relationship between geographical research and artistic practice. This might be characterised by the kind of material thinking that we witnessed at work with the use of objects and material practices Flora's studio, in the complex and sometimes conflicting imaginations of research and its role in the creative process that emerged from SOE. We also find a highly local and situated form of social research at work through Tania's mobile studio, recreated a new as a means through which she engages her site. Her studio is an assembly of people integral to the intensely local engagement of her work and thus to its politics. The coming together of geographical research and creative practices in the studio make clear the rich ways that research emerges as a set of material and immaterial practices. It becomes bound up with questions of what constitutes artistic labour, for the individual practitioner as well as the collective. It is also integral to how we conceive of relations between thinking and material making, whether concerning individual artists, of collaborative collectives, where metal workers and architects are brought together with art historians and theorists.

Notes

1 Virtual Field-visits to Oklahoma, the website has now been removed, being replaced by the studio's second home at University of Minnesota, https://www.geographystudio.org (accessed 5/2/2020).
2 Fraser, "The Studio"; Bauer, "From Bottega to Studio"; Jacob and Grabner, *The Studio Reader*; Buren, "The Function of the Studio"; Offman, *The Studio*.
3 Coles, *The Transdisciplinary Studio*.
4 Davidts and Paice, *The Fall of the Studio*.
5 Coles, *The Transdisciplinary Studio*.
6 Jones, *Machine in the Studio*; Bauer, "From Bottega to Studio."
7 Gotlieb, "Creation and Death."
8 Sjöholm, "The Art Studio"; Sjöholm, *Geographies of the Artist's Studio*; Norcliffe and Rendace, "New Geographies of Comic Book Production"; Leyshon, "The Software Slump"; Daniels, "Art Studio."

9 Bain, "Female Artistic Identity."

10 Leyshon, "The Software Slump?"; Warren and Gibson, *Surfing Places*.

11 Engelmann, "The Cosmological Aesthetics"; Jellis, "Reclaiming Experiment"; Sjöholm, *Geographies of the Artist's Studio*.

12 Engelmann, "Social Spiders"; Jellis, "Spatial Experiments."

13 These forms of work run throughout this volume, but are discussed especially in Chapter 4 on community.

14 Mel Bochner cited in Leaver-Yap, "Eva Hesse."

15 Leaver-Yap, "Eva Hesse."

16 Lippard and Chandler, "Dematerialisation of Art."

17 Bauer, "From Bottega to Studio."

18 Ibid.

19 Ibid.

20 Bauer, "From Bottega to Studio," p.182

21 Lippard and Chandler, "Dematerialisation of Art."

22 Ibid.

23 Cole and Pardo, "Inventions of the Studio."

24 Cole and Pardo, "Origins of the Studio."

25 Chapman, "The Imagined Studio."

26 Bauer, "From Bottega to Studio."

27 van den Akker, "Out of disegno."

28 Fer, "Eve Hesse."

29 Berger et al., *Eva Hesse*.

30 Fer, "Objects beyond Objecthood."

31 Fer, "Painting Drawing Sculpture."

32 Fer, "Objects beyond Objecthood"; Fer, "Studioworks."

33 Fer, "Studioworks."

34 I visited *In Real Life* on a number of occasions during its run in Autumn/Winter 2019–2020.

35 Ibid.

36 Jones, "The Server/user," https://www.olafureliasson.net (accessed 20/12/2019).

37 See for example the range of forms that geographer Sasha Engelmann has been involved in developing with the project Areocene https://aerocene.org/about_2020/ (accessed 20/1/2020).

38 Bishop, "Information Overload"; Rogoff, "Turning."

39 https://olafureliasson.net/studio (accessed 20/12/2019).

40 Jones, *Machine in the Studio*; Jones, "Event Horizon"; Jones, "The Server/user."

41 Engelmann, "Social Spiders"; Jellis, "Spatial Experiments"; Straughan, "A Touching Experiment"; Straughan and Dixon, "Rhythms and Mobility"; Woodward et al., "One Sinister Hurricane." https://studiotomassaraceno.org https://capefarewell.com http://www.symbiotica.uwa.edu.au (accessed 20/12/2019).

42 Coles, *The Transdisciplinary Studio*.

43 Ibid.

44 Eliasson, "Welcome."

45 Eliasson, "Welcome," p. 10.

46 Jones, "The Server/user."

47 Ibid.

48 See for example the interviews in SOE, *Open House*.

49 Eliasson, *Open House*.

50 Jones, "The Server/user."

51 Ibid.

52 Quotes in Coles, "Trandisciplinary Studio.'

53 SOE, *Open House*, p. 54.

54 Englemann, "Social Spiders"; Jellis, "Spatial Experiments."

55 See the range of Sasha's collaborations with Studio Saraceno on http://www.sashae-ngelmann.com (accessed 12/2/2020).
56 See the range of work on https://aerocene.org (accessed 12/2/2020).
57 Engelmann, "Of Spiders and Simulation," p. 310.
58 Engelmann, "Of Spiders and Simulation," p. 307.
59 I worked with Tania and her team in Tunis during October 2017.
60 http://antifestival.com/en/lebanese-artist-tania-el-khoury-announced-winner-of-anti-festival-international-prize-for-live-art/ (accessed 12/2/2020).
61 El Khoury, *The Audience Dug the Graves*.
62 https://taniaelkhoury.com/landing-page/; El Khoury, "Sexist and Racist People"; El Khoury and Pearson, "Two Live Artists" (accessed 12/2/2020).
63 https://2019.dreamcity.tn (accessed 12/2/2020).
64 https://taniaelkhoury.com/works/un-marry-us/ (accessed 12/2/2020).
65 Relyea, *Your Everyday Art World*; Gielen, "Community Art."
66 Ibid., Gielen, "Community Art,' p. 587.
67 Ibid., p. 581.
68 Jones, "raumexperimente."
69 Relyea, *Your Everyday Art World*, p. 592.
70 Hawkins, "Doing Gender and the GeoHumanities."
71 Doherty, *From Studio to Situation*.
72 Bond, "Enlightenment Geography in the Study"; Driver, *Geography Militant*.
73 Foster, "The Artist as Ethnographer"; Kwon, *One Place after Another*.
74 Ibid.
75 Demos, "Rethinking Site-Specificity"; Foster, "The Artist as Ethnographer."

3

LABORATORY

Geography is perhaps, as Scott Kirsch makes clear in his entry on laboratory/observatory for the Handbook of Geographical Knowledge, not best known as a laboratory science.[1] Yet geographers are important voices in debates about the spaces and practices of laboratories, historical and contemporary, and questions of laboratory geographies have long been important. Such geographies might map the modernist lab as a space apart, or as a site of blurred boundaries and boarders, between for example, landscapes and lab-scapes.[2] Most Geography schools or departments have labs, whether as sites of scientific experiment, or as the location for suites of computers for geovisualisation or remote sensing. Indeed, for many physical geographers the infrastructure of the lab is an integral part of their knowledge production. Further, for those who run schools and departments of geography there are often strategic reasons to emphasise its lab-based nature, with a range of benefits (as well as costs) accruing to subjects designated as 'science' or 'bench' subjects. Thus, while studies of geographers in the lab maybe far less rich than those of geographers in the field, laboratories are still an important part of geography, and as this chapter will explore, laboratory spaces, practices, and imaginations play an interesting role in relations between geographical research and creative practices.

Introducing an edited collection exploring 'new laboratories' Charlotte Klonk observes, that 'for well over a century now, the laboratory has been the epitome of creative experimentation, demonised and glorified as such in equal measure.'[3] As she continues, the laboratory has long been stylised as the *locus classicus* of the inquiring and inventive intellect.'[4] Discussions of the laboratory, like those of the studio, go back and forth with its demise and its changing forms. For a while actor network theorists were talking about the 'laboratization of society,' in which the extension of scientific infrastructure beyond the laboratory was understood to make society more like science. More recently however, Noortje Marres has been clear on how the opposite effects might equally arise.[5] In 1999, historians of science, Galison and Jones, could write of the demise of the laboratory as a fixed locus, suggesting that scientists increasingly pursue decentralised research projects where 'experiments are dispersed social-technical-spatial entities in which meaning is constructed at several peripheries.'[6] A decade or so later, however, Klonk and others, compiled an edited volume of laboratory architecture that suggests the resurgence of the research laboratory as a physical place.[7] A key driver for the enduring form of the physical laboratory building was the potential for communication and collaborations simply not achievable through dispersed virtual networks.[8] Laboratory architects Hawkins/Brown suggest,

> the key driver behind these buildings is collaboration... so you need to create an environment which allows [scientists] to do their own stuff, but then also come together. So while you make the offices and the labs as good as possible, the building is really all about the spaces in-between.[9]

The result of the back-and-forth discussions of the geography of the laboratory, has been a shift its imaginations. We now consider the laboratory as less the secluded 'space apart' of pre-war scientists 'who saw themselves as essentially solitary in their contribution with nature,' or the factory that evolved in the mid twentieth century, 'as scientists encountered patterns of work and places of produced altered by the vast factory production quotes of firms like Albert Kahn Associates and the Austin Company.'[10] Rather, the form and design of contemporary laboratories reference the 'totally upbeat state-of-the-art media and communications company,' like Google or Apple. This is not a one-way street. For the laboratory has emerged, in turn, as a force shaping creative organisations and institutions. Both London-based ArtsLab and the US-based Experiments in Art and Technology, founded in the mid-1960s, evidence the rich cross-fertilisation of ideas of the 'lab' (in the former's case with little sense of a scientific laboratory), as well as (in the latter case) a collaborative exchange of scientific practices with those of the studio. Indeed, the proliferation of 'lab' as part of the title of artistic spaces and organisations has been significant. Not least, as part of the characterisation Hans Ulrich Obrist and Barbara Vanderlind offered of the art world in their exhibition *Laboratorium*, in which they describe an art world constituted by 'networks, fluctuating between highly specialized work by scientists, artists, dancers, and writers.'[11] Perhaps the most intensified relationships between arts and the laboratory are found however in art-science collaborations. Indeed, the growth and importance of such comings together of artists and scientists in a range of geographical locations and across a range of disciplines should not be underestimated. There is a huge infrastructure of organisations that support and nurture these ways of working, of which geographers have made a number of studies.[12] These art-science collaborations are valued for the tools they offer art, for their critiques of science, as well as or the forms of communication they offer. Oftentimes their collaborative forms draw attention, the degree of their exchange or instrumentalisation of each other a subject of critique, as well as how together, they generate 'new' knowledge and practices.[13]

Physical geography laboratories have long been sites of artistic practice. The Leverhulme Artist-in-Residency scheme, discussed in Chapter 5, has funded a range of these collaborations, while other collaborations are funded through smaller schemes or emerge through informal connections.[14] The growing interest from physical geographers in creative practices is demonstrated by the evolution of the Visualising Geomorphology Group supported by the British Geomorphological Associated. This group was instigated by Heather Viles and Stephen Tooth and, as their exhibitions and a paper explores, are interested in a range of intersections between artists and geomorphologists, including, but not limited to communication.[15] Another emerging area of exchange is the cross-supervision of PhD students. Bethan Lloyd Worthington for example, at Royal Holloway University of London, is supervised by Quaternary scientist Danielle Schreve and myself. The funding opportunity for Bethan's studentship emerging partly from the work done in collaboration with Flora Parrott (detailed in the chapter on the field).[16] Whilst in 2017 Julian Ruddock completed a PhD between fine art and physical geography at University of Aberystwyth, supervised by Professor Stephen Tooth, a discussion of which is included below.[17]

In order to explore these intersections of art, geographical research, and the laboratory, this chapter falls into two key sections. In the first it explores those projects which have seen artists enter physical geographical laboratory spaces, whilst in the second it turns to consider the rise in the imaginations of the laboratory and its experimental practices within discourses and practices at the intersection of geography and creative practices. The empirical material for the first part of this chapter focusses on ethnographic work and discourse analysis of art-science collaborations based in geography and allied fields. This includes, principally, a nine-month artist residency at the WSL Institute for Landscape, Forest and Snow research outside Zurich where, in addition to following the artist in residence – Christina Della Guistina – through the inception, progress, and output of the residency, I was also a geographer in residence for a month, working alongside Christina.[18] Another important example is Julian Ruddock's doctoral work. This is detailed through Ruddock's PhD, in online sources, and has informed Tooth's ongoing work on visualising geomorphology. I also attended the exhibition of the work as the examiner for the PhD. A further core source for this chapter has been a series of documents and records (websites, blogs, calls for paper, as well as chapters and journal articles) that constitute the recent discourse around experimentalism and the creative lab space in the context of geographical research and creative practice relations.

Just as Peter Galison notes of the scientific laboratory, 'there can be no question of a single, unified history of the laboratory embracing all time and places' so is the case here.[19] Rather than a singular sense of the laboratory emerging from these forms of geographical research, I am interested to note how the laboratory and associated ideas of experiment have been shaped by, but also been put to work by these forms of creative research practice.

Artists and geographers in labs – new labs?

To get to Christina's studio you go through the main building, past the array of awards from key European Science Councils and historic displays recording contributions and discoveries made by staff at WSL. You pass the head and shoulders portraits of famous science staff, go around the side of the library, down the stairs past the microscopy lab, and around the corner. If you go up those stairs you get to the floor of offices where Christina has a space in a shared room full of computers – she goes there to print things and do work on the internet. Up the stairs again, and you get to the floors with all the other labs. We go down the stairs, press the bar on the door to exit the back of the building and cross the road, past the test gardens and into a flimsy, temporary building, a prefab with ill-fitting plastic doors and a hot plasticky smell. It is empty and has an abandoned feeling; the carpeted offices home to the odd office chair and table. Down the short windowless corridor, we come to the end, turn left and Christina's light-filled studio has taken over one of these office spaces. The windows that fill two walls overlook more test gardens full of saplings, another wall contains book shelves, whilst the far wall charts the development of her project. The wall is lined with paper and on it she has collaged scientific drawings printed and copied from books, sketches of trees drawn from life, plastic sample bags with small bits of twig and leaf, colour pictures printed from microscopes, and a range of graphs, notes, and computer print-outs. In the middle of the studio sit three long tables. One of them is to be mine. On the others are pieces of carved wood and cross-sections through tree trunks, leaves, piles of books about trees and tree science, heaps of print-outs of scientific articles. A microscope; a series of paints; and some pencils, paper, charcoal, and pens sit in various places on the tables.

It has been said, perhaps more in passing than with any degree of empirical discussion, that art-science collaborations are a site for the making of 'new laboratories.'[20] Further, as Beth Greenhough and others make clear, oftentimes interventions – in their case by social scientists or by animals – make these 'laboratory sites more complex, defined not only by their practices but by a whole series of other approaches which bring with them new means of entering and engaging with the field site/island laboratory.'[21] In what follows, I want to explore what kinds of laboratories emerged from two accounts of art and geographical science collaborations: the first a longer discussion of Christina's residency in the context of my ethnographic work on it, the second reflections on Julian Ruddock's exhibition and the accompanying written documents of the PhD and the exhibition catalogue.

Entering the lab?

Christina's residency was funded by the organisation Swiss artists-in-labs who have established a very popular and successful model for artistic residencies in science labs. Founded in 2003 and working out of Zurich University for the

Arts, and funded by a range of Swiss arts and science funders, the organisation works both in Switzerland and around the world to develop a model of art-science collaboration, based on intensive artistic residencies.[22] It has successfully exported this model to countries around the world, where its co-director Irene Hediger has overseen residency programmes in South Africa, Russia, China, and the Middle East amongst other locations. While in practice the programmes vary in length, their model tends to foreground intensity, ideally a nine-month span of artists working within the lab spaces. The residencies are competitively awarded through an open-call to propose projects for specific labs, with lab staff involved in the selection process. Residents are required to attend a series of symposia bringing together the artists and scientists to brainstorm and reflect on their experiences, and to offer 'crits' on their practice for the artists. Towards the end of the residency term Swiss AIL often conduct a process of reflective data collection. Film crews visit the laboratories and studios, and interview the artists and scientists involved, they also talk to lab directors. As both research process and outputs, these films have created the core material for three Artist-in-Labs books.[23] Interestingly, what emerges from these accounts is less stories of the final works and more accounts of work 'in process,' both art and science, and how these exchanges could bring about shifts, sometimes small, sometimes large, in artistic and scientific spaces and practices.

Christina Della Guistina spent nine months in residence at the Swiss Institute for Forest, Snow and Landscape research. WSL, as it is known, is an environmental science and physical geography lab complex on the outskirts of Zurich, best known for its contribution to identifying the ozone hole and their long-term permanent plot sites monitoring vegetation and soil health under climatic changes.[24] There was a dual meaning at work in how the scientists and eventually Christina came to speak of the 'lab.' On the one hand the lab was the complex of buildings, equipment, gardens, and plot sites – the latter many miles away in the Alps. On the other, the lab was also the research community within which Christina was embedded. This was a group of about twenty individuals who worked alongside each other sometimes on the same problems, but also ate lunch, played football, drove each other's children around, and attended BBQs and art openings together. During her residency, Christina lived close by, and would, like many of the scientists, go to the lab most days to work. She was originally given a desk in a postdoctoral study space with a computer and access to a printer and to a range of laboratory equipment, mainly microscopes, to help her develop her work. After a few days however she requested a studio space where she could experiment with the visual and sonic elements of her practice. After some negotiation, she was given a space in the empty prefab building near the test gardens. The space not only provided room for her to work through her ideas and make work, including three large artists' books and the sound piece, but also emerged as a site of exchange. She would invite the scientists to come and discuss the evolution of her works, the relationship between what she was doing and their scientific practices.

Over the course of Christina's residency, multiple sites around the laboratory complex become those of artistic making but also of display. Christina's working practices disrupt any sense of the linear understanding of the successive frames through which art or science might move – from studio to gallery to magazines, from field to lab to journal article.[25] Instead, we see Christina's artistic labour occurring across the studio and the laboratory, with her studio space set up in the laboratory complex. Her residency also saw her working in the office, in the field and in various different parts of the lab complex, the thin-section laboratory, the microscopy room, and the computer centre where all the real-time live data from the permanent plot sites in the Swiss countryside was being processed. She also produced a series of publically orientated works, engaging local communities near the field sites, and audiences in major European cities, with WSL's science practices. Exploring the idea of the laboratory that emerges from Christina's work is to see perhaps less a 'new' form of laboratory, but rather a contribution to the challenges those studying science have made to the idea of the modern lab as 'a space apart.' This was a space for science whose practical needs, geographic location, and popular imaginary resulted in a separate space needed for the evolution of the universal facts of a place-less science.[26] Yet, as Beth Greenhough observes, attention to science-space relations through concepts, such as risk, uncertainty, and the ideas of experimental processes, has brought forth the lab as a site whose boundaries are constantly being challenged.[27] Far from spaces of clear boundaries, structures, and coherence, laboratories have emerged from geographical work on experiments with islands, with rewilding and with household microbiomes and so on, as 'spaces of disruption, contingency and risk.'[28] Furthermore, these are spaces whose boundaries are constantly being re-negotiated by their human and non-human inhabitants, including ancient breeds of Longhorn cattle, water voles and infectious diseases, and here, I would argue artists.[29]

To elaborate, Christina's studio was not just a studio folded within a lab, but was also a site of encounter between the scientists and the artwork that was evolving in relation to their work. This occurred through Christina's presence in their research group meetings, in various open studio events she put on to coincide with their own public facing 'open-lab' programme, but also through a series of lunchtime workshops. One lunchtime, for example, scientists from various parts of the lab came to visit the studio set in their grounds. Entering gingerly at first, they gradually relaxed as they recognised familiar materials – graphs from their work, microscopy that they work with daily, traditional specimen drawing which they observed they had not encountered since they were students, and photographs of their field sites. They handled specimens carefully but with a kind of familiarity – they pointed to and discussed diagrams amongst themselves. We sat and listen to a series of recordings – part of Christina's evolving sound work. What, they ask, are we listening too, the transposed recordings Christina replies, what does this mean, they ask? Christina switches to Swiss German, the lab's *lingua franca*, and explains: she has taken the data streams from the monitoring stations at the lab's permanent plot sites and has transposed them into a sonic score. The continual flow of measurements for carbon dioxide, water vapour, temperature, gases exchange, and so on have been sonified; we are 'listening' to the processes of transpiration, photosynthesis, and the wider hydrological cycle. The assembled group were not entirely sure what to say; 'it does not sound like trees,' one observed. Others agreed, perhaps it should sound more-watery, more gassy. It is odd conversation in which the imagination of a process, insensible to humans, a process with no human-based sonic register, is judged through wider cultural imaginations of soundings of air, water, and plants. After listening to the score and noting the changes they would like us to explore, we turned to another aspect of the project, focussed on looking, a more familiar mode of knowing. Christina drew over the microscopic images, and they gathered around them to talk. There is discussion of staining techniques, questions about how she has been attaching her cameras to the microscopes, discussions of the biology – do stomata open and close gradually, and we just have not seen it yet, or is it a binary state – open or closed? Christina is experimenting with time lapse as a possible way to visualise this process. They are intrigued, could she, an artist, do this? They are open to discussion, seeming to enjoy musing on how fast or slow this might happen, the different scales and staining techniques that might be needed. They are hopeful it might work; it would be an important contribution if it could be recorded.

Christina's creation of a studio space in the midst of the lab, a space that was both a site for making and for display, also created a site for exchange, in which different ways of thinking about the objects of science emerged. The environmental scientists – from Botanists to Hydrologists – worked with the streams of data and images every day, but in Christina's studio they were brought to these objects and data in new ways. What if we listened to the streams of figures that recorded the changes in gas and water content of leaves and stems? What do those trees we have been working on for years sound like? Why do we think they sound like that? What does it even mean to talk of trees making sounds in this way? Some of these were serious questions, some facetious, asked half in jest. This was less about doubting Christina's work, or making fun of it, but seemed instead a tentative exploration of what this might mean for their own practices. Across the group, and during the residency we witnessed a going back and forth with how to know these trees and their processes, the many ways to document, engage, and record them, including these imaginations of the sonic properties of trees, shaped by science and aesthetics. Reflecting on the coming together of different knowledge forms in his studio, Eliasson describes,

> a back and forth between looking at something for yourself, by yourself, pondering the issues, and then looking at it from the plural point of views of the team, of friends, knowing from where they speak, knowing and evaluating through their glasses with what they see, its very inspiring. A kind of shared felt, looking develops.[30]

In the case of Christina's work, it felt less like a shared sensing was emerging, an in-common attunement to the objects of artistic research and scientific study, the trees, the gasses, the hydrological cycle. Rather, it seemed enough that the coming together of the different ways of knowing these shared objects offered the chance to appreciate what each might offer differently. I want to take up these ideas further, in considering Julian Ruddock's display of science.

On display

The exhibition is held in the University Art Centre's enviable gallery space. It is a space designed for the display of art rather than a requisitioned and refitted seminar room, as some university exhibition spaces are. The space is black, and the lighting has been carefully designed to draw us to the works on show (Figure 3.1). The exhibition seems to have five 'zones' of sorts. I was immediately drawn to the core, lying horizontal, lit behind a glass window, samples drilled at regular intervals from its sides. I wondered whether this is no longer needed, on loan, will it still be used for science in the future or is it merely being preserved now? Drawn around the corner by light and sound, I find the width of the gallery and the back third of its length given over to a darkened space, with a large projection filling the end wall. I sit on the floor leaning against a wall, and watch a core move vertically up the screen, taking us backward in time, 500,000 years deep by the end. The film is 24 hours long, I stay and watch for 24 minutes, thinking this seems like a nice symmetrical amount of time. I try to work out how many thousands of years I have seen pass by, it is roughly 8,000 years. I move around the three individual darkened booths, where I am encouraged to sit and watch the films. The monitors are full of talking heads, of landscapes and shots of science labs. The first film is about fieldwork in the dried lake basin of Chew Bahir, Southern Ethiopia where the core has come from. The landscape and indigenous people are the stars of the film, the potential significance of the landscape and the lake in the history of evolution explained. I try to stop myself wondering about ethics forms and safeguarding and keep my focus on the aesthetics of the film and its content. The second film is about the theory and concepts behind the science of The Hominin Sites and Paleolakes Drilling Project (HSPDP), the third features the international research team, and its network of labs and research facilities that have supported the analysis and visualisation of the core. The final two films in particular feel, because as I sit in the chair I face the individuals talking, a bit like a one-sided conversation. I exit the booths, and move around to the large graphic display panels, 'hypothesis,' 'fieldwork,' 'laboratory': these read. Images and text offer context for the science, its processes, findings and fieldwork — laboratory is illustrated by a large-scale picture of a diatom. I finally enter the little room I had seen on entering the exhibition. The temperature changes, it is a warmer, lighter, there is a slightly earthy smell. I look into the plastic tray in front of me. It is mounted on black fabric, the mud inside it is cracking, a landscape in miniature, the textures of the sediments standing out in places, in others still a bit muddy looking. The play of scales is intriguing. It is somehow both an epic landscape from which these cores might be taken, but also a microscale site of the processes of deposition and drying that created the core and the animation in miniature (Figure 3.1).

Field Notes from an exhibition visit, June 2017

FIGURE 3.1 Exhibition view, *2A Earth Core: The Hominin Project*. (Image, Julian Ruddock)

The exhibition *2A Earth Core: The Hominin Project* is one of the seven elements of practice Julian Ruddock presented in his practice-based PhD.[31] The written document offers auto–ethnographic reflections on the use and interpretation of scientific data, material, and images in fine art practices around climate change. Over the course of the discussion we are introduced to the working processes behind a series of forms of practice. These are three large-scale bodies of art-work: *The Dyfi Project* (exploring geomorphology in the Dyfi River catchment), *The Explorer Series*, and the *2A Earth Core: Hominin Project*. We are also interested to two exhibitions – *Predicting a Climate Archive* (2012) and *2A Earth Core: The Hominin Project* (Figure 3.2) – and two important high profile symposia – 'Future Climate Dialogues' and 'Strata'.[32] Across this body of work multiple laboratories are evoked. This includes the visualisation lab central to the Dyfi river project, which made extensive use of Lidar data; the high-tech science laboratories; and institutions that analysed the cores, the gallery, and seminar room as 'labs' of a form as well as the physical space and community in Geography and Earth Sciences at Aberystwyth University, with whom Julian collaborated to produce the sculptural landscape.

Each project emerges from Ruddock's accounts as an intersection of lab, field, studio, archive, and gallery space in their creation and display. Walking through the exhibition, the sites of science become folded with those of art making: the field, the scientific centre and institution, the painting and sound studios, the laboratory as well as the whole scientific community, and extended engagements with indigenous communities whose lands and experiences were involved in the making of the work. The installation of the core sets a 'scientific' tone for the exhibition. This was not the core collected and worked on by the team, but another one, some 40,000 years old, collected during a previous trip. Lit from above, behind glass, the effect is of a rare artefact, an object of science to be valued. Indeed, walking through the exhibition is almost a journey through the process of scientific knowledge making: hypothesis, laboratory, and field. As Julian notes when talking about the graphic posters, he was concerned that they situated the team's research as proper science, that the artistic context not challenge the authenticity of the research practice. Discussing the making of the animation for 'Earth Core: The Hominin Project' Julian presents this as a process of negotiation of the practices of the laboratory and his aesthetic concerns. He recounts an earlier iteration of the piece in which he had chosen images of 'best' parts of the core – 'best' here being those he found most aesthetically pleasing. These were spliced together to form the film. Reflecting on this work, and talking to the scientists involved, he became unhappy with his aesthetic editing process, and decided to use images of every bit of the core, and also to include the meta-data at the sides of the images – those tags and markers he had original cut out. These choices are framed in discussion as a kind of fidelity, a sense of an authentic treatment of the practices and products of the laboratory. The sense of science and of the laboratory that emerges is of a complex intellectual and logistical enterprise, to which the international movement of samples, equipment, personnel, and data are key, and whose focus is the 'production and validation of new empirical knowledges'.[33]

FIGURE 3.2 Julian Ruddock, *Earth Core*. (Image, Julian Ruddock)

A rather different sense of the laboratory emerges from the making of *We come from this place* (Figure 3.3). With its own enclosed space, the trough of mud, light harshly from above by staged lights had blossomed deep cracks. In the thesis and the catalogue, this work is described as a collaboration between Julian, and two scientists in Aberystwyth's Geography and Earth Science's Department: Henry Lamb and Helen Roberts. It uses fifty kilograms of discarded material from the scientific process. This material was potentially half a million years old, but was so mingled with material from the sides of the core as it was extracted that it is useless for dating. Ruddock recounts how this discarded material was shipped from France to Wales and become the focus of the artwork in what the artist describes as the most successful of his collaborations. This sense of success seems to stem from the collaborative involvement of the scientists in developing the form of the piece.

The account of the work's production fascinatingly entwines artistic and scientific inspirations and experiments. Alongside records of the discussion and testing of scientific equations to get the right mix of material and water, are notations of similar artistic works. These include Robert Rauschenberg's *Mud Muse* (1968–1971), Goldsworthy's *Clay Room* (2007), Alice Alycock's *Clay Number 2* (1969), and Walter de Maria's *Earth Room* (1977–ongoing). Questions of the aesthetics of presentation also become questions of slurry consistency, drying time, a monitoring process that involved the wider lab community, including students, as well as debates about how to source the right tray – in terms of both size and look. Ruddock reflects too on how the installation of the sculpture in the exhibition created a space that looked more like an experiment than he had originally intended;

FIGURE 3.3 Julian Ruddock, *We Come from This Place*. (Image, author's own)

We decided to leave the lights drying the mud for the first week running up to the opening, and used barrier tape to stop visitors accessing the space where the hot lights were positioned. I very much liked the look of the space at this point as it had the feel of an experiment, which indeed was the case.[34]

The 'new labs' that emerge from these art science collaborations resonate with those studies of the laboratory which challenge the modern sense of it as a space apart. Instead, these are sites in the midst of myriad interactions with other sites of science and of art – from the field, to the test garden, the permanent plot, the dry lake bed, the indigenous community, the water flume, and so on. The labs of these art-science collaborations are far from 'formed, defined and waiting,' but rather, as Greenhough and others writing on laboratory practices suggest, emerge through those practices, in this case artistic practices, that challenge their boundaries and put at risk stable fixed understandings of their objects of enquiry, practices, and technologies of research and the certainty of their outcomes.

Laboratory imaginations: experimenting with creative practices

If geography has been understood less as a discipline of the lab than of the field, it is interesting to observe the growth of a laboratory imagination and associated ideas of experimentation, as important force in discourses and practices at the intersection of creative practices and research. If the sway of these ideas is indicated through the creation and naming of a series of collectives and groups in these terms – SenseLab, The Office of Experiments and the Deep Space Laboratory (both run by Neal White), the Temporal School of Experimental Geography, and The Experimental Geography Studio – it is also signalled through the rise in the idea of the experiment.[35] Experiment has become a common idea, both across geography more generally, as well as a series of commentators have noted, in the context of the relations between creative practices and geographical research, through what has been termed 'experimental creative geographies.' That interdisciplinary imaginations might take up the lab and the creative experiment as a site has a long history, rooted most specifically in Modernism. We might think, for example, of Walter Gropius, the German architect, describing the modernist arts and crafts workshops at the pre World-War one Bauhaus as laboratories. The Bauhaus was an interdisciplinary school of arts and crafts whose combination of theory and practice was orientated towards the production of beautiful and useable industrially produced objects for the home, but whose methods and practise came to shape interdisciplinary education (especially in the arts) across Europe.[36] Whilst in 1939, Alfred H Barr, Founding Director of the Museum of Modern Art in New York, declared the museum a laboratory for the arts.[37] We might think too of how John Dewey famously reclaimed the idea of the experimentalism for avant-garde arts practices, noting how experimental dispositions could be found across aesthetic and political practices not just within science. In the context of these long standing discussions, I want, in this section to take a closer look at imaginations of the laboratory and of the experiment that emerge at the interface of geographical research and creative practices.

Experimental geographies

In amongst the range of terms that have emerged to denote the relationship between creative practices and geographical research, the idea of 'experimental geographies' or 'experimenting with geography' is one with significant traction.[38] These sometimes 'aesthetic,' 'creative,' or 'artful' experiments sit alongside wider and diverse geographies of experiment that emerged in the early years of the twenty-first century in the context of a wider ethos of experiment across social science.[39] This 'experimental turn' or 'injunction to experiment' has a series of forms which between them rethink the experiment as an empirical occurrence and as form of knowledge.[40] Thus creative geographical experiments join a longer list of experimental forms that includes 'experimental society'; an 'ethos' or 'disposition of experimentation'; the collective experimentation of hybrid collectives which redistribute expertise amongst scientists, stakeholders, and publics; and the human and non-human relations of 'cosmopolitical experiments.'[41] Many celebrate how such experimental richness heeds the long-standing concerns of historians of science not to oversimplify the experiment into a caricature. This includes what Gailson observes as the deplorably 'impoverished representation of experiment' or what Lorimer and Driessen describe as 'the singular understanding of an Experiment, associated with positivist approaches to natural science.'[42] Yet others caution, perhaps not surprisingly given the richness of this list, that accompanying this work has been a call for a closer parsing of these forms of experimentalism. Thomas Jellis, in a wonderful body of work, calls for a concern with the 'specificity of the experiment' whilst Jamie Lorimer and Clemens Driesson worry about a 'lazy lexicon,' and Richard Powell and Alex Vasudevan warn against a 'slack metaphorics.'[43] In response Angela Last enacts the experiment as a 'travelling concept,' possessing 'a range of orientations and sensibilities when put to work in different areas of enquiry.'[44] As such, I want to ask what aspects of the experiment are 'put to work' in the context of the coming together of geographical research and creative practice?

Sources for reflecting on this question include discussions in interviews, and research papers that enact these experiments, as well as PhD theses by Thomas Jellis and Dominic Walker.[45] A further important source has been a series of blogs, often now not updated, but which for a few years in the early 2000s formed an important point of diffusion for creative experimental geographies.[46] The popularity of blogs for those working in these ways is interesting, and perhaps relates not only to changing modes of academic communication, but also to the demands of these creative experiments. For, as the discussion will explore, these experiments celebrate the ongoing and provisional over the singular finished output, and a blog's immediacy and transiency perhaps suits the research-in-process feel of these experiments?

One of the 'logics' (to borrow a phrase from Thomas Jellis) of creative experiments found across these accounts is a diverse and often distributed sense of creativity at work. We see, for example, the artist as the instigator of the experiment, we see creative practitioners enrolled in collective experiments with others including scientists and social scientists (on which more below) and we also see the use of creative practices as a means to take account of redistributed expertise, valuing the views and experiences of various publics.[47] Thus, perhaps unsurprisingly, a further logic of these experiments is that of the collective or of collaboration. Creative practices offer powerful means to assemble collectives or 'hybrid fora.'[48] People are brought together through, for example participatory art around environmental problems, or around issues of ecological restoration or urban inequality.[49] Collaboration and collectively has emerged as a means to cope with uncertainty and with complex problems that require multiple solutions but also to 'encourage communities to learn from such unknowns and to live and blossom with them.'[50]

Another logic that clearly unites many of these forms of creative experiment concerns the generative, exploratory form of experiments – they focus on world crafting or recomposing rather than testing or description. These are experiments that have been loosed from any logic of repeated testing of a hypothesis or to confirm results.[51] Instead the ethos of these experiments appears as generative, as forward thinking, as productive. Indeed, whether we look to McCormack's experiments with dancing, moving bodies, to David Pinder's or Cecilie Sachs Olsen's performances of the arts of urban exploration as experiments with everyday urban spaces and politics, or Sasha Engelmann's discussion of Tomas Saraceno's atmospheric experiments, we find experiments creating 'new' worlds.[52] These are experiments that seek less to describe and know, as to offer geography 'the means to map out new potentialities for being, doing and thinking.'

A final logic of experiments I want to draw out concerns their resistance to the singular and accepted form, what could perhaps be called the commitment to the production of novelty. This is a sense of experiment that refuses to be confined by the caricature of the experiment of the natural sciences, but is rather a function of another methodological stance. If the idea of 'experiment' is generally recognised as being generative of new ways of thinking and doing geography, of which creative practices have formed an important subset, then what I take from across these diverse discussions of experiment (creative or otherwise) is their shared querying of what methods are and do. Akin to the fieldworkings that emerged from the discussion in Chapter 1, these experiments tend not to be about the conduct of *a priori* methods and protocols. As such, the 'experimental' of creative experimental geographies directs us, usefully I think, towards a rethinking of creativity, that is less the deployment of specific 'creative' practices (however skilfully or amateurly), and is instead a considered sense of a disposition towards processes of ongoing transformations. In other words, these are methods responsive to the world they engage, they are transformed, and transform it, as they go. To explore this further I want to turn to the example of Montreal Sense-Lab, a lab explicitly created at the intersection of creative practice and research, and a site that has been influential for geographical thinking.

Creative 'Labs'

If there is a diffuse sense of experiment at work across the intersections of creative practices and geographical research, there is also a more concentrated sense of experiment and the lab to be found in a series of temporary and permanent spaces – whether buildings, rooms, relational groups, or residency-like experiences – that call themselves laboratories. In discussion with creative practitioners who had experienced spaces/events that summoned up the image of the laboratory, it was interesting to explore the ideas of the laboratory that were emerging. These were clearly quite some distance from that sense of the lab as a space of structure, of coherence or fixity, and as a space of the rigid reproduction of scientific method. Instead, the imaginary conjured a sense of process, exploration, and intensive experimentation. As one artist–PhD student noted,

> the laboratory, ah, a capsule, a time or a freeing space. I would love to be part of a lab because you can experiment, and today we are in an art economy quite ruled by entrepreneurship and branding and strategy, so quite economically driven. So the academic process and the PhD proposes space to be able to practice, and sometimes that is right, but sometimes there are too many forced goals, and a lab is like a sort of intense, focused mode, but not, importantly an accelerated mode of production. The Lab is like, umm, a capsule, yes, a zone of experimentation, we are not allowed to experiment any more, we have to produce.

Fascinatingly, this seems to return us to that sense of a laboratory as a space apart. Importantly, though, not a space apart geographically, necessarily, but a space that enables distance from a certain focus on production and output, a space that enables open-ended experimentation. Another, expands on this sense of the lab and experimentation, noting that the sort of artist labs she had been at – often time-bound experiences, like residencies she explains – were also spaces that enabled failure. Labs, she explained, are important 'safe spaces for failure'; they are spaces that 'enable sitting with processes, with exploration and probing'; failure, she continues, is vital for good work to develop. To be always focussed on 'productive success' does not let anything really good emerge she reflects.

This sense of process, experimentation, and failure seems to characterise many of those spaces that have come to call themselves 'lab.' To explore these ideas in more detail I want to spend some time reading accounts of the Montreal Sense-Lab as a site that specifically describes itself as operating at the interface of creative practice and research.[53] The SenseLab was founded in 2004 by Erin Manning, a Professor of Relational Art and Philosophy in the Faculty of Fine Arts at Concordia University (Canada). It has already attracted the attention of a range of geographers including Thomas Jellis and Derek McCormack, both of whom have visited, participated, and written on SenseLab.[54] Jellis and McCormack's accounts are amongst a series of papers, books, PhD theses, and blog posts that reflect on what happens within the lab. Indeed, Manning describes the Lab's core outputs as its process, and a third of her book with Canadian philosopher and social theorist Brian Massumi offers an account of the lab's development.[55] As such, there is rich material through which to explore SenseLab and to engage with basic questions around what constitutes the 'lab,' what happens 'at/in' the lab, and to further reflection on how it frames its own processes as outputs.

SenseLab describes itself as a 'laboratory for thought in motion,' what does this mean? As the website suggests, it is

> an international network of artists and academics, writers and makers, from a wide diversity of fields, working together at the crossroads of philosophy, art and activism. Participants are held together by affinity rather than by any structure of membership or institutional hierarchy. The SenseLab's event-based projects are collectively self-organising. Their aim is to experiment with creative techniques for thought in the act....The SenseLab has adopted the term 'research-creation' to describe its activities, with the goal of fundamentally rethinking "theory" and "practice" in a way that overcomes the all-too-common antagonism between the two.[56]

It is very clear that this is not a laboratory in the bricks and mortar sense. Nor it is an organisational structure akin to research groups, although it could be considered to be a continually reassembling network, conducting experiments with various forms of props and communication devices. Yet the lab also refuses organisational forms and structures in fixed terms. As Manning explains,

> Dedicated as it was to the practice of the event, the SenseLab avoided defining itself into a formal organisational structure. It was conceived as a flexible meeting ground whose organisational form would arise as a function of its projects, and change as the projects evolved... membership would be based on elective affinities. Anyone who considered himself or herself a member was one. The result was a shifting mix of students and professors, theorists and practitioners, from a wide range of disciplines and practices.[57]

As such the lab is perhaps best understood through what it does. A section of Manning and Massumi's book "Thought in the Act" entitled "Propositions" uses the device of twenty propositions as a means to narrate the intersections of the practices and philosophies of the lab and in doing so evolves its origin story.[58] This story seems related as much to their reading of process philosophies – Deleuze and

Guattari, Whitehead and others – as it does in a critique of the neo-liberalisation of research, the coming together of research and shifting forms of the creative and knowledge economy, and the effects of this on the professionalisation of artists and the rise of artists in the academy.[59] Of particular interest to Manning and Massumi is the emergence in 2003 in Canada of a new funding category 'research-creation' which they suggest offers 'hybrid forms of activity promising to capture for research the creative energies of artists working within the academic institution.'[60] For SHHRC, research creation is understood as

> Any research activity or approach to research that forms an essential part of a creative process or artistic discipline and that directly fosters the creation of literary/artistic works. The research must address clear research questions, offer theoretical contextualization within the relevant field or fields of literary/artistic inquiry, and present a well-considered methodological approach. Both the research and the resulting literary/artistic works must meet peer standards of excellence and be suitable for publication, public performance or viewing.[61]

SenseLab's "Proposition 0" is to practice 'immanent critique.' In other words, they sought to recognise that as university-based researchers/artists they were not outside these contexts of funding councils and the neoliberalisation of the creative economy and of academy. Instead, they desired to 'work in the thick of tensions – creative, institutional, urban, economic – and build out from them'. Indeed, one of the goals of SenseLab is to 'curdle research-creations annexation to the creative economy.'[62] One of the clearest indications they give of what they mean by this occurs when they narrate their practice as one of almost time travel. They focus they suggest, on that hyphen between research and creation that goes back in time to 'unwind the divergence of research and creative practice and to undo the institutional structures that try to keep them apart.'[63] They seek a kind of 'origin' of thought and practice that appreciates how 'making is already thinking in action,' and 'conceptualisation a practice in its own right.'[64] The two they suggested merge in 'technique,' which they define as 'the engagement with the modalities of expression a practice invents for itself.'[65] As they explain elsewhere 'to dance: a thinking in movement,' 'to paint: a thinking through colour.'[66] Such meetings in 'technique' they emphasise also needed to be 'constitutively open-ended,' 'experimental emergence effects of an ongoing process.'[67]

As their propositions progress, the 'lab' emerges as site which generates experiments with bodies, with texts, with string, and twine, tents, mangos, squashes (the vegetable), art galleries, streets, and alley ways.[68] The list could go on, but the primary device for these experiments is the event. It feels tempting to try to flesh out, to detail these experiments, but it is hard. Many of the accounts, whilst seeming to narrate the process, are all rather abstract; little real detail and few images are offered, and as such, I find it hard to conjure up a sense of what exactly a day spent experimenting with SenseLab might involve.[69] Perhaps more important is their understanding of the event as narrated powerfully through an account of a conversation with philosopher of science Isabelle Stengers, when she indicated that her criteria for attending an academic event were that it be just that, 'an event.'[70] As they continue, an event in these terms was, 'a collective thinking process being enacted that can give rise to new thoughts through the interaction on site.'[71] The forms of event Manning and Massumi's propositions detail congeal around accounts of shared virtual and in-person reading groups, designed to multiply perspectives rather than lead to singular accounts (although Jellis reflects on the success of this when the force of their presence tends to shape discussion).[72] They also offer accounts of camping, cooking, and bodily movement, of 'relays' between reading practices and 'movement experimentations, writing activations, materials-used explorations, and spatial/environmental creations'. There is also a fair amount of manifesto writing, with further propositions, injunctions, and so on. As much, it seems, as these event-experiments are focussed on questions around 'thought in the act,' about bodies, movement, and philosophy, they are as much experiments in/with the form of the events themselves. In other words, a core business of the lab seems to be a sequential set of experiments, complete with failures, each of which informs the next. Together they pose many possible answers to questions such as, 'what new forms of collaborative interaction' does this model of research-creation imply, 'what conditions' are necessary, 'what does it mean to organise for emergence?' Further, we might ask, what are the implications of this experimental emergent-event orientation for established forms of interaction of creative practice and research, such as the conference paper, the seminar or the gallery exhibition?[73]

Whilst committed to facing outward from academia, especially through a new participatory and community-based three ecologies lab, theirs is also a laboratory whose origin, framing, and practice is constituted through and by the process philosophies read at many events and more subtly by the framing of the wider discussions through the vocabularies of these philosophies.[74] As a result, the discourses about the laboratory and its activities which circulate within and beyond geography (including in this text) often reproduce these forms of philosophy – not least because they offer such a key to understanding how the laboratory works. The effect of this framing is to draw a perhaps unhelpful distinction between the SenseLab's practice and those of wider intersections of creative practices and research. Fascinatingly, the more I read across their events and practices, the more I found similarities between their descriptions and events I attended across the art world, and took to be a natural part of how that world unfolded. Further, I found in their accounts similar approaches to those I was learning about through my collaborations with Flora. This included the use of objects to encourage a thinking and practicing together, the setting of organisational conditions that attempt to disrupt hierarchies and roles people easily fall into through things like seating patterns and how they introduce themselves, and the importance of food. I had been struggling with how to think through these ways of working, in fact with even identifying some of them as repeated patterns in events. So SenseLab's detailing and thinking-practicing of its form of lab experimentation offered me a new lens through which to explore these activities and strategies – what they might call techniques. This resonates with that sense of the experiment that emerged earlier. The creative experiment as less a set of fixed practices, using set methods, than a disposition towards the world as it unfolds in process.

New laboratories?

If art-science collaborations are sometimes claimed to create 'new laboratory' spaces it is interesting to reflect on the ideas of the laboratory that emerge at the intersection of geographical research and creative practices. As this discussion has suggested, in line with wider geographical writings on laboratories, these are perhaps less 'new laboratories' than a reinforcement that laboratory spaces are not the closed-off space apart that our modern imaginations of them might suggest. Instead artistic work in laboratories might be understood as part of a wider set of practices from beyond the lab, through which the laboratory emerges as a complex site, defined, as Greenhough suggests, 'not only by their practices, but by a whole series of other approaches which bring with them new means of entering and engaging with the field site/island laboratory.'[75] Artistic engagements with the laboratory suggest both a respect for the knowledge making practices found there (witness Ruddock's concern with getting the science right), but also a desire to multiply understandings of scientific objects, practices, and spaces. This includes the idea of the experiment. For while, the 'injunction to experiment', as Jellis puts it, across social science of late might be associated with the persistence of positivism, it seems that those working with ideas of creative experiment, owe much to the wider understanding of experiment as never solely this singular scientific caricature, but rather also a distributed field of practice long associated with 'avant-garde' arts. Whilst experiment is, as Last and others remind us, a nomadic concept whose iterations need careful parsing, the logics of the experiment of creative experimental geographies directs us to rethink creativity less as a set of specific practices – whether this be painting, writing, or dancing, and is instead a considered sense of a disposition towards processes of ongoing transformation.

Notes

1 Kirsch, "Laboratory/Observatory."
2 Galison, "Three Laboratories"; Kohler, *Landscapes*; Greenhough, "Tales of an Island-Laboratory."
3 Klonk, *New Laboratories*.
4 Ibid., p. 9.
5 Marres, "As ANT is getting undone."
6 Gailson and Jones, "Factory, Laboratory, Studio," p. 527.
7 Klonk, *New Laboratories*.
8 Galison, "Buildings"; Galison and Thompson, *The Architecture of Science*.

9 Galison and Jones, "Factory, Laboratory, Studio," p. 524; Galison, "Three Laboratories."

10 Galison, "Three Laboratories."

11 Obrist and Vanderlinden, *Laboratorium*, p. 584.

12 Straughan and Dixon, "Rhythms and mobilities"; Straughan, "A Touching Experiment."

13 Woodward et al., "One Sinister Hurricane."

14 See for example, http://geohumanitiesforum.org/news-announcing-the-creating-earth-futures-commissions-2018/ (accessed12/2/2020); Alice Ladenberg and Alexander Shekin, University of Oxford, Environmental Change Institute, http://www.alice-ladenburg.com (accessed 12/2/2020); Overend et al., "The Bones."

15 https://www.geomorphology.org.uk/working-groups/visualising-geomorphology_(accessed 12/2/2020); Tooth et al., "Visualising Geomorphology"; Dixon et al., Art-Full."

16 http://bethanlloydworthington.com (accessed 12/2/2020).

17 Ruddock, "Navigating the Uncertainties," https://cargocollective.com/artscience climatechange (accessed 12/2/2020).

18 Fieldwork at WSL conducted March–December, 2011. I continued to work with Christina through a grant funded by the Swiss National Science Foundation to explore the potential of pop-up lab/exhibition spaces.

19 Galison, "Buildings."

20 Klonk, *New Laboratories.*

21 Greenhough, "Tales of an Island-Laboratory."

22 http://artistsinlabs.ch (accessed 12/2/2020).

23 Scott, "Artists-in-Labs"; Scott, "Networking in the Margins"; Hediger and Scott, "Recomposing Art." and Science.

24 https://www.wsl.ch/en/index.html (accessed 20/1/2020).

25 Buren, "The Function of the Studio."

26 Galison, "Three Laboratories"; Knorr-Cetina, *Epistemic Cultures.*

27 Greenhough, "Tales of an Island-Laboratory."

28 Ibid., p. 225.

29 Greenhough, "Tales of an Island-Laboratory"; Hincliffe et al., "Urban Wild Things"; Overend and Lorimer, "Wild Performatives"; Lorimer and Driessen, "Wild Experiments"; Lorimer et al. "Making the Microbiome," p. 225.

30 Eliasson, *Open House.*

31 Ruddock, *Navigating the Uncertainty.*

32 Rich accounts of these projects are found on https://cargocollective.com/artscience-climatechange (accessed 20/1/2020).

33 Thrift, *Spatial Formations*, p. 43.

34 Ruddock, *Navigating the Uncertainty*, p. 195.

35 Office of Experiments, http://o-o-e.org/fieldwork/; Deep Field Lab, http://o-o-e.org/fieldwork/blog/2019/07/30/deep-field-part-1-initial-conditions/; Temporal School of Experimental Geography, https://tsoeg.org (accessed 12/2/2020).

36 https://www.bauhaus.de/en/das_bauhaus/45_unterricht/ (accessed 12/2/2020).

37 https://www.moma.org/momaorg/shared/pdfs/docs/press_archives/4249/releases/MOMA_1969_Jan-June_0082_56.pdf (accessed 12/2/2020).

38 Kullman, "Geographies of Experiment."

39 Last, "Experimental Geographies"; Davies, "Where do?"; Enigbokan and Patchett, "Speaking with Specters"; Ingram, 'Experimental Geopolitics"; McCormack, "Thinking in Transition."

40 Powell and Vasudevan, "Geographies of Experiment"; Jellis, "Reclaiming Experiment."

41 Latham and Condradson, "The Possibilities of Performance"; Hincliffe et al. "Urban Wild Things," p. 644; Whatmore, "Mapping Knowledge Controversies."

42 Kohler, "Lab History"; Lorimer and Driessen, "Wild Experiments," p. 170.

43 Jellis, "Reclaiming Experiment"; Lorimer and Driessen, "Wild Experiments"; Powell and Vasudevan, "Geographies of Experiment," p. 1791.

44 Last, "Experimental Geographies," p. 880.

45 Jellis, "Reclaiming Experiment."
46 Blogs include: Experimental Research Network, http://experimentalnetwork.org (accessed 12/2/2020); as well as those associated with to those of the individual practitioners, such as Merle Patchett's Experimental Geographies in Practice, https://merlepatchett.wordpress.com (accessed 12/2/2020).
47 Caprotti and Cowley, "Interrogating Urban Experiments"; Hincliffe et al., "Urban Wild Things."
48 Whatmore and Landstrom, "Flood Apprentices."
49 See for example discussions in Ingram, "Washing Urban Water"; and Hawkins et al., "The Arts of Socioecological Transformation."
50 Gross, *Ignorance and Surprise*, p. 34.
51 Marres, "Testing Powers of Engagement."
52 McCormack, *Refrains*; Pinder, *Visions of the City*; Engelmann, The Cosmological Aesthetics; Last, "Experimental Geographies"; Kullman, "Geographies of Experiment."
53 https://www.senselab.ca (accessed 12/2/2020).
54 Jellis, "Reclaiming Experiment"; Manning and Massumi, *Thought in the Act*.
55 https://senselab.ca/wp2/inflexions/ (accessed 12/2/2020); Manning and Massumi, *Thought in the Act*, Intermediations book series: https://senselab.ca/wp2/book-series/ (accessed 12/2/2020).
56 https://www.senselab.ca (accessed 12/2/2020).
57 Manning and Massumi, *Thought in the Act*, p. 90.
58 Manning and Massumi, *Thought in the Act*.
59 https://www.senselab.ca (accessed 12/2/2020); and Manning and Massumi, *Thought in the Act*.
60 Manning and Massumi, *Thought in the Act*, p. 85.
61 http://www.sshrc-crsh.gc.ca (accessed 12/2/2020).
62 Manning and Massumi, *Thought in the Act*, p. 87.
63 Manning and Massumi, *Thought in the Act*.
64 Ibid., p. 89.
65 Manning and Massumi, *Thought in the Act*, p. 85.
66 Manning and Massumi, *Thought in the Act*, p. VIII.
67 Manning and Massumi, *Thought in the Act*, p. 89.
68 Manning and Massumi, *Thought in the Act*.
69 Ibid.
70 Manning and Massumi, *Thought in the Act*, p. 90.
71 Ibid.
72 Jellis, "Reclaiming Experiment."
73 Ibid.
74 https://senselab.ca/wp2/3-ecologies/3-ecologies-institute/ (accessed 12/2/2020).
75 Greenhough, "Tales of an Island-Laboratories."

4

COMMUNITY

The sun is going down over the rooftops of Tunis. Tinny music is playing out of the speakers. We are dancing. I watch as Tania, the artist I am working with, joins hands with a group of women from teenagers to those many decades older, laughing and cajoling them she gets them to start moving to the beat. She is a good dancer, some of them join in willingly, others have to be encouraged, I join hands too. Some, like me, are stiff and self-conscious, no idea how to move, others are happy in their sensuousness laughing, shedding their shoes, dancing on their toes, shimming their hips, reminiscing about past weddings, trying to teach their friends, and me, how to make the traditional moves. They collapse laughing and pointing at me, grasping my hips and shoulders trying to make me relax, shifting my position. The music continues, the call to prayer starts, we keep dancing. Someone produces scarves, we start dancing with them wound around our bodies, held over our heads. The light fades, we are laughing and tired.

We move into the room off the balcony, we sit down with scripts, and one of the women starts reading; her voice is strong in the empty room; the story is one of abuse, of escape, and of finding refuge. She speaks of the women's shelter, where we are working, the only one in Tunisia, its location a secret, the safety found here, the friends to be made. One after another, the women read the stories they have been assigned, some in strong clear voices, heads held high, others speak more softly, stumbling, eyes turned down, heads bowed.

After a few hours we leave the shelter and wind our way through the streets of the Old Medina. The entrance to the courtyard of the beautiful Dar is through a small door set into a larger door, we lift our feet high to get over the larger outer frame with its ornate carvings, inches thick. Standing in the centre of the courtyard we look up at the layered balconies and floors of the old guest house. This will be the setting where the women will tell the stories, stories that interweave personal, intimate geographies of violence and refuge, and the intricacies of everyday spaces and practices of safety in the city (rooftop escapes and secret signals) with the Old Medina's long history as a site of shelter and safety from persecution.

Field-notes from research with Tania El Khoury in Tunis, Autumn 2017

'Community' is a vibrant site for the intersection of creative practices and geographical research. Here we find everything from geographers enrolling creative practices (from song-writing to story-telling and theatre practices) as part of their participatory action research, to collaborations with community-based arts organisations. We also find artists, like Tania El Khoury, a description of whose work *UnMarry Us* opened this section, coming to do geographical research on and through site-specific live arts practices, engaging in-depth with the context as well as the people.[1] We might think too of geographical involvement in the production of public art forms, as well as the enrolment of creative practices within geography's wider agendas of public intellectual work and stakeholder engagement, what in the UK falls under the 'impact agenda.'[2]

Foregrounding community, as a site for the intersection of geographical research and creative practices, offers the chance to explore the potential of rich exchanges between the practices and literatures across the diversity of participatory art and participatory geography. While there are many points of difference, crucially what both share are the challenges they pose to normative practices and forms of research and of art. If the practices and literatures of participatory art and participatory geography have tended to slip past one another, then the lens of community brings into view a series of illuminating intersections around process and product, around issues of judgement – whither aesthetics, asks one art theorist, concerned by the slip into sociological discourse of these practices – and around the ongoing challenges of negotiating art world and academic institutions. Ahead of addressing these three points of intersection more directly, I want to detail further these two fields – participatory geographies and participatory art – and the shared challenges they mount to 'normative' practices of art and research.

The expanded field

There are a growing number of projects, like Tania's, that are situated at the intersection of geographical research and site-specific live art and participatory practices. Sometimes, these projects are brought about by the individual artist (often with a support team), and other times the projects are the result of collectives, such as Cecilie Sachs Olsen's work with zURBS, or the Digital Media Collective Furtherfield, both of whose work will be discussed further in what follows. The projects also cover a huge range of mediums, including live art, textiles, theatre, and dance-based practices. The emergence of these forms of art making – in which people are both the medium and form of work – was one of the more controversial evolutions of twentieth-century art. It was also one of its most geographically pervasive, especially under political shifts in arts and social funding, often exported from Western art funding contexts and city governance practices. These projects can now be found in locations around the world – from the standard bastions of Western art in the UK, Europe, and America to work in China, India, and a range of Middle Eastern countries.[3] The terminology for these practices is varied – community art, participatory art, experimental communities, dialogic art, littoral art, relational aesthetics, contextual art, socially engaged art – and the theoretical and art historical trajectories equally so, and often conflicting.[4] They range from very local, community-based arts practices, seen as marginal to the 'art world,' to the high-profile, controversial discussions of 'relational aesthetics,' and the revival of debates about politics and aesthetics driven through and by continental philosophers such as Jacques Rancière.[5] The forms of participation within these projects vary. Sometimes communities are enrolled throughout the process of art making, from conception, to the contexts of display. In other projects, an art activity – craft projects or gardening (for example) – is designed ahead of the participant's engagement, and their role is to carry out this activity. Sometimes a participant's role is simply to co-produce materials with the artist, materials that are later used within or come to inspire art made elsewhere. Importantly for many much of the art writing produced on these ways of working, participatory arts are also shaped by a hugely diverse understandings of 'the social' and the ways communities might be engaged, including 'dialogue, collaboration, process, diversified audiences, democratic practices.'[6] Yet it is this sense of social relations as a medium that holds together these diverse ways of work, and offers the locus for their critical force with respect to the practices and political economies of the art world. As such, these art forms have made a central contribution to twentieth century's ongoing reframing of art as object or artefact, to the understanding of an exhibition, and to the figure of the artist and their relationship to the institutions that support them (patrons, galleries, funders) and 'audiences.'[7]

To elaborate: if we take the practice of geographer – artist Cecilie Sachs Olsen for instance – her work is produced as part of a collective, zURBS.[8] Like many such collectives, the goal of the work is not the production of a singular artefact valued for its rarity but, rather, a series of situations that are produced in a sequence of on-going projects; an array of social events, workshops, performances, seminars, and publications. Such shifts in the form of art also, as Claire Bishop's book on the subject makes clear, shifts the idea of audience.[9] Continuing twentieth-century art's breakdown of the audience as viewer, as a distanced beholder of the work, zURBS practices cast the 'audience' as participants within and co-producers of the work. It is perhaps no surprise, then, that participatory art is challenging for the art world's galleries and institutions, as well as for commercial dealers. Based in social relations and events, it is hard to document effectively, difficult to market (often practitioners resist it being advertised as 'art' in case participants are put off), and challenging to display let alone sell – not least as there is no object or artefact to monetise. Despite, or perhaps because of, these challenges participatory arts practices have come to occupy a prominent place in the critical art world as well as the public sector. They are, for example, common forms of public commission, which are often funded as government-supported arts-based social projects and now make up the back-bone of many biennials and politically themed exhibitions. Interestingly, in her discussion of participatory art Bishop stops short of including what she writes as 'transdisciplinary, research-based, activist or interventionist art, in part because these projects do not primarily involve people as the medium or material of their work.'[10] In doing so however she misses out on those forms of participatory research, by geographers and others, that proceed through creative practice. As such, the chance to bring together understandings of participation within art with those from other disciplines is frustratingly missed.

It is interesting that participatory geographies, including those that enrol creative practices, raise similar questions for geographic research as participatory art raised for the art world. These include the impact of shifting forms of research production, the nature of 'outputs' and diverse relationships with participants. Indeed, at the heart of most participatory research is the ethos, Louise Waite and Cath Conn explain to

> encapsulate a desire to move away from 'traditional' qualitative research where a researcher elicits information about 'them' in an extractive manner in favour of interactive research with an avowed commitment to the co-production of knowledge between the researchers and the researched.[11]

Further, participatory research is often also driven by an explicit quest for research to operate as a vehicle for social change, asserting that the commitment to the processual nature of participatory methods is an integral way to bring about this change. A rich range of creative practices – from visual methods of video making, drawing and photography to story-telling and theatre making – have formed valuable ways for participatory geographers to connect and work with communities.[12] Sometimes geographers might collaborate with a creative practitioner and a community, throughout the whole project – from aims to methods, analysis, and outputs; other times creative practice might offer one technique amongst many used with participants to evolve ideas and even to collect data.[13] The logics and framings for the use of creative practices within participatory geographers are varied. They include, their potential to help express traumatic and often unspeakable things, in other words to access certain kinds of 'data' and the value these practices offer for 'connecting communities' around issues of concern or community futures (from hydrological issues to gender-based violence).[14] A number of geographers have found in creative practices the means to help deconstruct the dominant discourses and social hierarchies, and forge the kinds of horizontal power relations they seek with their participants.[15] Further, creative practices are not just valued as research methods but also as valuable forms of communication within and beyond the communities they work with.[16] Not only are arguments made about how these practices can bring new voices – often unheard – to ongoing conversations, often with big stakeholders (such as planners, politicians, or corporate interests), but it has been argued that these forms of work can also assemble new publics around their chosen issues.[17] While much is shared across these comings together of participation, geographical research, and art, they should not, of course, be collapsed together, and in what follows discussion of how the fields of participatory art and participatory creative geographies engage with three themes – process, judgement, and negotiation – proceed as much from differences as from similarities.

Process is more important than output – or is it?

Along the top of the opening screen of the website 'Play South Westminster,' a cartoon airplane (drawn by Oliver, 36, South Westminster) pulls a hand-written banner, reading, 'What will you Save?' A player, jumping from one platform to the next, can choose from a series of games, each of which encourages an engagement with one of the many social challenges facing the South Westminster community. The virtual world of this city-ward, conjured by its diverse community in drawn and painted form and replicated here in co-produced video games, is one where Boris Johnson, portrayed as a Machiavellian Mayor of London, rides his community bike high above the city streets. On these same streets, caricatured figures of smartly dressed 'gentrifiers' are urged to 'grab the cash… be evil-kill community spirit, push up rents, hike house prices and force out families.' Elsewhere in this virtual city space, a small mouse dashes around platforms/streets armed with a fire hose trying to put out the burning buildings, the last line of defence after the closing of fire stations across London. The cute and whimsical world of South Westminster the community portrayed in their games engages us with some of the lived experiences of this city ward. For, if one view of South Westminster focusses on the millionaire's mansions, leafy parks and international art galleries for which the area is famous, these games use their whimsy to encourage us to look beyond this to see a story of worsening inequality, a decline in local community and a loss of services under government austerity (Figure 4.1).

Reflections on *Play Your Place*[18]

FIGURE 4.1 Screen shots from *Play South Westminster* game. (Image, author's own)

Whenever I talk about the *Play Your Place* project developed by digital media arts collective Furtherfield, I am always a little stuck. Should I foreground the video games it produced as visual artefacts or discuss what happened over the course of the participatory processes of drawing, making, and gaming that are the heart of the project and the means by which the video games are produced? I am further torn because despite valuing the local participatory processes, I am also very aware of how for all sorts of reasons, for the Furtherfield team, for the participants, and for myself, the display of the video games in the context of the Tate Britain's *Folk Art* (2014) exhibition, seen by millions of people, was important to celebrate.[19] The local workshops were held with residents in the context of a local summer festival 'South West Fest,' held in South Westminster London (Figure 4.2). South Westminster is one of the UK's most socially diverse electoral wards, containing both some of the nation's richest individuals and its seat of power, but the ward also has one of the country's highest percentage of rough sleepers, and a considerable proportion of the area's housing stock is social housing. Tate Britain is set in the midst of South Westminster, yet many of its residents have never entered the gallery space, despite its free collections. Thus there was certain politics at work in the situation of the community-generated video games in the context of the gallery's summer blockbuster exhibition on folk art. The games were displayed, and could be played, alongside a detailing of the process through which they were produced: the original drawings made during the local festival, the videos of the discussions residents had with us about the drawings, and the workshops in the Tate's digital studio that developed the games from those drawings. There were also a series of networked computers that enabled visitors to view and interact with the open-access platform onto which all the game elements had been uploaded, theoretically allowing everyone, anywhere, to play their own place. This intersection with the Tate, and in particular the inclusion of the work within the *Folk Art* exhibition, was an important marker of esteem for the project and its team. It was also an interesting moment for the community to assemble in an art world space which many of them might not normally frequent, despite it being around the corner from many of their homes.

What both 'good' (a term considered later!) participatory geography and participatory art share is at least a notionary privileging of process over product. In other words, what happens in the 'doing' of these works is asserted as being as important, if not more important than their ostensible 'outputs' – journal articles, data sets, works for exhibition. Indeed, for many, what happens and is learnt in the doing is the primary output of these ways of working. Yet of course, within both art worlds and the academic world of geography, there is a challenge posed by the 'outputs' needed and required of those involved in making these works. Participatory geographers have reflected for example, on the challenging intersection of participatory research with normative academic knowledge production practices, whether of journal article production or the institutionalisation of the need for academic research to create 'impact' beyond academia.[20] The focus of participatory researchers on local, in-depth community engagement throughout the research process – from research questions through to dissemination of results – seems to be at odds with the apparent retrenchment of an elite politics of knowledge production across academia, not least in its assessment of how we value the difference research might make beyond the academy.[21]

These are not dissimilar challenges to those faced by practitioners in the art world. As Claire Bishop writes, 'today's participatory art is often at pains to emphasise process over a definitive image, concept or object. It tends to value what is invisible: a group dynamic, a social situation, a change of energy, a raised consciousness.'[22] The stakes are high for those creating these so-called dematerialised works. There are implications for how they find presence in galleries and exhibitions; how they are approached and analysed by historians and critics, and thus how they enter a 'canon,' securing the artist's reputation; as well as how they engage audiences. The latter involving questions of both how they reach and recruit initial participants, as well as how they find meaningful ways to communicate their activities to later audiences for the work.[23] There is, then, an often necessary reconciliation between the dematerialised social process and the object (its circulation on the art market, its preservation in the artist's archive or in the pages of volumes of art history, or its record in the journal article by the artist-academic). Further, for those practitioners who wish to engage in-depth and over the longer term with local communities, this can be at odds with the temporalities of funding regimes and art world circuits of display, which, as discussed in Chapter 2 on studios, can favour patterns of mobility linked to the globalised circuits of the art world and its festivals.

FIGURE 4.2 The *Play Your Place* stand at South West Fest 2014. (Image, author's own)

As local community members walk around the park where the festival stands have been set up, a number stop at the Play Your Place booth. The stand has been arranged with a series of seats at folding picnic tables to create drawing stations, each station with ready access to paper, paints, and pencils. As they sit down participants are asked to think about an issue or a challenge they think their community faces. Included at the drawing stations are some prompts to get people thinking. Written on fluorescent card shapes – a speech bubble or a spikey edged cloud form – are prompts: 'place,' 'actions,' 'beings and things'; 'the bigger picture' to help direct participants' attention. They offer a basic 'scaffolding,' encouraging discussion of what to draw and about the elements of local experience to explore further. Sitting around the table in the park in South Westminster, sometimes sheltering from the rain, people began to talk, sharing paint-pots and coloured pens as well as stories, discussing their lives and experiences through their drawings.

The written accounts we have produced of *Play Your Place* have tended to, in the end, foreground what happened in the process of the work. In this we are inspired by both accounts from art history and theory and from participatory research, which tend to emphasise the need to evidence claims around what these creative practices do. In other words, to flesh out the form and kinds of the relations that are produced through these works, rather than just accept them as happening and as an unqualifiable good. In trying to make sense of what was going on, we draw on those geographical accounts of creative practices and participation that place emphasis on 'the interaction of individuals and their own personal growth that comes about during the process of production.'[24] In other words, 'it is not the material outcome (that is, the film) which makes participatory video workshops worthwhile, but the process that leads to it.'[25] This emphasis on what creative processes do is made very clear in geographers Rachel Pain and Kye Askins's discussion of their project with the African Community Advice North East (ACANE) centre in Newcastle Upon Tyne in northeast England. ACANE is a refugee-led community organisation that aims to support African asylum seekers and refugees through a philosophy of whole-community integration.[26] The art project that Pain and Askins initiated involved different approaches to art making carried out over several months with a group of twenty-one young people aged between five and fifteen years old from both African refugee and white British-born backgrounds. They describe how 'these paintings, and more importantly the processes and discussion that went into producing them, began to identify points of similarity between the young people from different backgrounds.'[27] They are careful not to suggest that these were always positive social interactions, as they note:

> It is critical that we emphasise there were negative encounters as well as positive ones, and often it was hard to separate these out: they were not always separate... the divisive social relations that were problematic for young people in their everyday lives were fully present on site. Exclusions were evident: at times we were drawn into negotiating between conflicting views and behaviours, never as neutral observers but as actors within

the group dynamics...The key point is that, while the project involved young people exploring their differences, similarities were also realised and enacted through the material objects.[28]

As this project demonstrates, artistic practices can enable participatory research – understood here to offer the potential for cross-cultural exchange and for generating greater cultural understanding and social cohesion. This is achieved, not only through creating sites of encounter and dialogue, but also by 'creating space for embodied, multilingual, marginalised experiences to be expressed in visual form.'[29] Different creative practices enable different kinds of things. Drawing on art history and theory to consider *Play Your Place*, we queried what the medium 'drawing' did. As well as offering participants a 'ticket to talk' to each other and to us, we also found in the process a reshaping of the places, times, and paces of research. So while our initial questioning 'what problems do you have with this area,' or 'what change would you like to see around you,' might result in a puzzled look, the on-the-spot nature of talk-based methods shifted as scribbles became doodles and as fresh paper was reached for. As outlines were sketched, and colour was shaded, discussions swelled, sometimes expressed during the drawing process and for others expressed later, to camera, as we asked participants to describe their pictures.

Yet however much the local-community-based drawing process was valued, the process was also one in which there was a clear need to manage a set of anxieties about the transition from the drawings in the local festival to workshops in the digital suite at Tate Britain and thus to the *Folk Art* exhibition space. In contrast to the large numbers of locally generated drawings produced during the festival, the workshops to formulate those drawings into video games drew a more select and specialist crowd, many of whom were students, some passing gallery visitors. Whilst some worked on computers, others of us sat around tables discussing the key themes emerging from the drawings, watching back the video showcases where local residents discussed their drawings with us. I was struck by the similarities between our processes and those of photo-voice or elicitation methods, where participant-produced content – images, a video – becomes a site of focussed discussion. Eventually key scenarios were chosen and game images selected. As we discussed the game scenario, how a desire for more dog parks or the closure of fire stations and other vital services might feature in the games, we edited the images and generated new ones to fill the gaps. We created 'rewards,' discussed what 'winning' would look like (the ironies of collecting coins in a game critiquing gentrification) and formulated ideas for how you could 'loose a life' (failing to build sufficient community spirit?). In these discussions further details of the challenges and experiences of lives-lived in South Westminster became apparent. The images were then uploaded onto the *Play Your Place* platform and mock-ups of the games where produced and played by visitors and by us on the big screens in the digital studio. Over the next few days, guided by local community members and Tate visitors the suite of games emerged.

The processes of devising these games created a dialogue between the games on display in the main gallery of the Tate, the digital studio (on the ground floor of the Tate) where we had devised them, and the drawing and video stalls at the community festival around the corner. Across these sites it became clear that rather than everything leading up towards the display of the finished games, this was in fact part of an on-going process of exhibition. This was further extended when we consider the uploading of the finished games and the various game elements – sounds, characters, backdrops – onto the *Play Your Place* platform to be played as they were or remixed into other local games.

The process felt like an 'experiment whose outcome was altogether uncertain'; we were never sure what, if anything, was going to manifest for the *Folk Art* exhibition and in the process had to accept the different festivals, workshops, and exhibitions as incidences of the wider project's 'different phases of its successive materialisations.' For me this was hard to get comfortable with, yet the Furtherfield team were more used to how participatory practices tend to be sites of ongoing production, requiring as Bishop suggests a rethinking of the exhibition as something other than a collection of already made art objects.[30]

This ongoing enfolding of process and product that *Play Your Place* exemplifies is similar to that found within participatory geographer's accounts of their use of artistic, video, and photography methods. Whilst early accounts of these practices tended to foreground the importance of the creative practices as part of the process, more recent accounts have begun to grapple with the tensions between process and output. Reflecting on participatory video methods with domestic violence victims in Cambodia, for example, Katherine Brickell explores the importance of collective editing and community pride in displaying and discussing their work. Indeed, while the focus of accounts of participatory research might be on the process, myriad demands for the outputs of this work emerge: whether it be as prompt for further discussion; as a site for communication of the research; or, in terms of the journal articles through which both these projects are being consumed, as a written marker for the outcome of the research and its circulation. Divya Tolia-Kelly, for example, captures a common sentiment when she reflects on the 'unique communicative and social power that the arts can exert within the public sphere.'[31] There is, for sure, no denying the ways that many of those working at the intersection of geographical research and participatory arts identify its value not only for working with participants, but also for aiding in the communication of their voices/ views to wider groups.

Bringing together participatory art and participatory geographies results in the emergence of nuanced discussions about relations between process and output. It is of course important to value process, but for good reason practitioners can and do recognise and value 'outputs' (artworks in exhibitions, journal articles) as important for circulating ideas, gaining funding, building reputation. What emerges too is a shift in how we understand outputs, such that, for example, rather than a finished presentation of works in an exhibition, work might be considered as 'always in display' during the process, as it becomes a focus of discussion amongst participants. What follows builds on these discussions, revis-iting Tania's work before looping back around to *Play Your Place* to reflect on the challenges around judgement and value that these works pose.

Whither aesthetics – the slide towards sociology?

We sat on the roof of the Dar, and looked over at the opposite roof the other side of the court-yard, where three of the participating women/performers were slowly moving to the music, dancing with their wraps. One by one, they got up, and moved across the walkway towards our roof space. As they entered, they moved to the centre and started speaking. The first person to speak was too quiet, and could not look at the small audience, the second was fine and as her section progressed was building a good audience rapport and had remembered the choreographed movements that had been designed collaboratively during the last rehearsal. A third, however, could not remember her lines and stood there awkwardly trying to recall the movements. They were all frustrated, we were frustrated. The performance would not work like this, the audience would not be able to grasp what was going on, the stories were important, the immersive effect of the performance would be ruined, the message would not be conveyed. But these women were not trained performers, by choice and/or by culture many had little public exposure, and few had any experience speaking in public let alone performing or dancing. The rehearsal ended, there were some difficult discussions amongst the artist and her team about whether the performance would be good enough, how to recon-cile the desire for the performance to be good for the performers, who were clearly increasingly nervous about letting everyone down, with the challenges around their lack of experience and their nerves. I left Tunis the following day, pondering as I travelled back to London how the dynamics of the work would be resolved.

A few days later I heard from Tania, she had tried something new. What she mused, if the performance became a kind of exploration of the building. How about rather than the fo-cus being on the roof, what if we moved the performance up through the building, how about if rather than perform as separate from the audience, the performers created more intimate spaces with fewer people, where they could talk more naturally if they wanted, or perform, whatever suited their style. Some chose to tell stories in whispers to one or two audience members, others chose to distribute food as in a traditional wedding feast and share stories over the shared food, others danced and invited audience members to dance with them, some told stories while painting traditional henna wedding patterns on the audiences' hands.

Field-notes, Tunisia, Autumn, 2017

Site-specific and participatory art, as well as participatory geographies have long offered the kinds of practices that jam a spanner in the works of forms of judge-ment and critique associated with the art world and with geographical research. Some of these issues played out very clearly during the making of *UnMarry Us*. Core to the piece was the dynamic and presence of the women from the shelter, yet of course few of them had any experience of being the focus of attention, let alone performing for audiences, all had however volunteered to be part of the piece. As Tania and the team wrestled with how to evolve the work, discussion centred on questions of performer experience as set against the aesthetic effect of the finished piece. Tania was clear that the performers were key; their experience, comfort, and safety was paramount, but of course, the piece needed to be right aesthetically, and it needed to work, not least for the performers themselves.

The evolution of *UnMarry Us* dramatises some of the ongoing concerns art critics and theorists have around these forms of art, where normative conditions through which art practices have been judged – primarily in terms of aesthetics, and a sense in which art might extend the work's avant-garde cutting-edge discussions – are disrupted by the terms and forms of these ways of working. Claire Bishop goes so far, somewhat disparagingly to ask, 'whither aesthetics?', seeking to challenge what she sees as a 'slide into sociological discourse.'[32] It was these questions of how to judge the so-called 'social turn' in art, as well as how to study it that perhaps confirmed its place within art history and theory. Two vocal figures in these debates have been Claire Bishop and Grant Kester – and while they often come to blows – both agree that these art forms shift understandings of art and thus the terms of art history and critique.[33] In the introduction to her 2012 book length study of participation, *Artificial Hells*, Bishop observes the challenges she faced in studying these time and place-bound event-based works and in grappling with their social dimensions. Whilst in a reflective article for the on-line art journal *e-flux*, Grant Kester argues that the "October" model of art criticism, named after the *October* journal responsible for introducing critical theory – most often post-modernism and post-structuralism – to art critique, was no longer valid.[34] Both ponder the new modes of art historical research these works demand from their critics, no longer will viewing images or archival work be enough, nor will the abstract distance of theory (perhaps a caricature?) be appropriate. Instead, in-person models of fieldwork are needed. For Bishop, a 'sociological model' premised on several site visits 'preferably spread out over time' was the answer, whilst for Kester, an anthropological model of fieldwork akin to ethnography was forwarded as the solution.[35] Both, however, find the questions of critique even more thorny and challenging than questions of the research methods by which to approach these forms of art making. Bishop, for example, worries over what she sees as an uneven inclination towards the social component of this project, suggesting that contemporary art's 'social turn' not only designates an orientation towards concrete goals in art, but also the critical perception that these are more substantial, 'real' and thus more important than artistic experiences.[36] She is worried at the foregrounding of questions of what 'social work' art does, concerned not only with the fixation on the primary achievements of art as being those of empowerment, of collectivity, or of agency but also with the devaluation of aesthetic force and expressiveness in the light of these social goods. She also reflects on how the 'perceived social achievements' of these forms of work are never 'compared with actual (and innovative) social projects taking place *outside* the realm of art.'[37] As such she continues, they remain uncritically considered, working at the level of 'an emblematic ideal,' deriving their critical value from their opposition to 'more traditional, expressive and object-based modes of artistic practice' rather than any achievement of social goals.[38] Perhaps the epitome of this is found in curator and critic Nicholas Bourriaud's notion of 'relational aesthetics,' a form of aesthetics based on social relations, rather than conceiving of these social relations as apart from aesthetic questions as Bishop appears to do.[39]

If we look to discussions of participatory geographical research and creative practice relations, we see, in fact sociology and aesthetics, to use Bishop's shorthand do not need to sit in opposition, although at times it might appear that they do.[40] Within participatory geography one school of thought foregrounds social process; observing, for example, how, within participatory video, 'it doesn't [and shouldn't] matter whether the image shakes and the takes are very long. What is important is the social and cultural dynamic associated with this image'; technical and artistic grounds for selection were seemingly backgrounded.[41] Indeed Rachael Pain and Kye Askins offer a fascinatingly detailed account of negotiating these concerns in their ARCANE project.[42] Working with stakeholders, they found the first of their two projects with young people foregrounded co-production with the group, but produced work that stakeholders deemed not high quality enough for display. Yet when, for the second part of the project they used an artist to create 'higher quality' artefacts, more palatable for the stakeholders, they found this reduced the positive experiences for the group. They reflected that this seemed to, 'foreclose the new emergent social relations' they had witnessed in the first section of their project. Yet, as they are careful to observe, it was pointed out that had they used a different kind of artist these two goals might not have seemed antithetical.

If we look at a cross-section of projects we see that indeed, for many working at the intersection of geographical research and art, the need to balance aesthetics and social work is crucial. As the discussion around the quality of the performance in Tania's work opening this section suggests, to be participating in a creative practice and to feel oneself to be 'not good enough,' can also be a barrier to participation and disempower participants, which is inappropriate and unfair. This was also at work in the context of the drawing events associated with *Play Your Place*. One of the first to sit down at the drawing stations was a young boy, together with his parents. As he started to draw, they started to talk. They mention the lack of open space, the boy wants a dog, and there is a discussion about how parks are disappearing (leaving less space to walk dogs). During their discussion, the generic background image of green space and sketched buildings evolves into a park for a dog. As soon as the boy starts to draw the dog however, he gets visibly frustrated; his dog does not look like a dog. The process of discussion breaks down, the focus is on the quality of the dog, its tail, the scale of its legs, the dog is almost scribbled out, but one of the facilitators steps in and encouragingly helps refashion body and tail.

Participant drawing is often used in situations where there is an assumption that participants will struggle to express ideas. This might involve young people or research contexts where there are cultural and linguistic differences, or it might concern things not easily put into words – expressive practices, or lived experiences for example. Across such discussions there seems to be an apparent assumption that drawing is somehow a more basic form of communication than words, and is somehow universal and accessible. It is, of course, important not to overlook issues raised by participants around not being good at drawing. Indeed, it has been noted that literate children often perceive an artistic inadequacy, principally worried (like the boy with his dog) that their drawings do not look right.[43] Furthermore, while drawing is often assumed to foster a certain form of attentiveness to that being drawn, frustrations at not being able to draw can break down that relationship.[44] As such, the qualities of judgement extend also to the participant's reflections on their own practices.

Others have suggested that participant judgement of aesthetic quality also effects how participants relate to the work, and hence its social effects. Tolia-Kelly, reflecting on her experience of working with artist Melanie Carvello argues for the importance of high-quality professional artists in collaborations. She observes how it 'enabled rigour in producing visual materials that are socially and culturally recognised as "Art", and provided essential advice and skills necessary to avoid the risk of participants viewing the research process as naive, "experimental", unethical or patronising.'[45] Collaboration with a professional can work in other ways too. Hester Parr, for example, describes how working together with participants to learn new skills of video-making was a great leveller of the kind of power relations between research and participants, that participatory research is seeking. Cecilie Sachs Olsen offers another dimension to these discussions in her discussion of the practices of the urban art collective she works with, zURBS.[46] Cecile describes a continual practice of negotiation between ensuring, on the one hand, that the work feels accessible, and on the other that participants know it is polished and 'proper' not just thrown together, she reflects;

> If looks like something we pulled together quickly, our DIY aesthetics, because we want it to look like that. We want it to look unfinished. We want that because then the barrier for them taking part in changing things is lower of course than if it feels really polished. But obviously, also then if something doesn't work for them, people think we did not think it through. We did, and we spend, like years developing these concepts, appearing effortless is much harder.[47]

It is not just the participants whose engagement with the work might be effected by the aesthetic qualities of the piece, but also the audiences who might be otherwise engaged by the outputs. The quality and appeal of the creative product does matter, by creating 'more potential for community audience engagement and/or community pride in the representations developed and shared.'[48] This is clearly the case when arguments are being made about the communicative value of the outputs and the potential they have to assemble new publics for the issues they take up.[49]

It is not just the value and importance of aesthetic judgements of these works that is of interest, but also questions of how to assess their 'social' work.[50] So while Bishop can reflect that very rarely are participatory works judged against social projects from other sectors this is not always the case. Indeed, it is rarely the case in participatory research, for while they might not always be judged against social projects, they are held to high standards around participatory ethics. Questions are asked for example, around the quality of the experience for the participants, the nature of participation and its effects are pressing, and indeed, given the community buy-in needed can often make the differences between the realisation or not of these projects. Such evaluations are still tricky, for as Pain and Askins observe, 'any "gold standard" of participation as an equable sharing in knowledge production across the whole course of the process rarely works out in predictable ways.'[51] Furthermore, there are also clear examples where participatory artists are held to task for the kinds of relations they build with communities. We find, for example, drama and performance scholar-practitioner Jen Harvie and performance theorist Shannon Jackson asking pointed questions about the 'social work' of these forms of art and about the labour politics of these ways of working.[52] In a very different artistic context art theorist Hal Foster holds artists to the standards of good ethnography, calling them out for a nonchalant execution of methods with significant political implications.[53] He is scathing of how artists, unaware of the detail of ethnographic methods or their complex politics histories of othering and cultural extraction, are too hasty in their desire to mobilise the aura and authority that surrounds these methods. This withering critique is not dissimilar to those mounted of site-specific art, where rather than get to know a place, an artist simply reproduces the same method in a series of locations. Such work, as explored in Chapter 2 on studios, is often seen as a production mode driven by the circuit of biennales where artists are required to produce new locally engaged work for a series of sites in quick succession. The terms of critique that emerge here include, geographic specificity (but not parochialism), appropriate forms of social connection /relations (between artist and site, and between participants at that site), but also that there should be a high-quality aesthetic output as a result of the participatory process (rather than that process as the high-quality output) in order to facilitate communication and the raising of the needs or profile of the participants.

Clearly there are likely complex and competing needs to be negotiated and it is perhaps unlikely that all these different forms and terms of judgement could be responded to at once. This is, of course, perhaps a particular challenge for those working at the intersection of geographical research and art whose work straddles both geography and the art world, having to respond to the terms of judgement emerging from both. While these discussions of judgement will be taken up further in the final three chapters of the text, to close this one, I want to turn to consider negotiation as a key characteristic emerging from these forms of practice.

Negotiation: instrumentalisation

One strain of Modern Art theory situated, and sought to preserve, art as a form apart, characterised by autonomy from the everyday world, a transcendence from daily life. Indeed, Modernist art critic Clement Greenberg sought to define artistic media through their resistance to assimilation into the cultural industry. In 1960 Greenberg wrote, 'the arts could save themselves from [post-Enlightenment]levelling down only by demonstrating that the kind of experience they provided was valuable in its own right and not to be obtained by any other kind of activity.'[54] If, as has just been explored, questions of how we judge, and how we critique, art have become one point at which we need to reflect on the 'social entrenchment of art,' then in what follows I want to muse on how the valorisation and practice of the arts of entanglement, rather than those of transcendence, brings with it complex questions of negotiation, especially institutional negotiation. Whilst the focus will remain on community based and participatory practices, the arguments advanced are relevant across the coming together of creative practice and geographical research. Not least because, as discussion will suggest, universities – as institutions – are a key point of such entanglements, and as accounts have made clear, everything from their politics of labour and outputs to their temporalities, require negotiation. Indeed, questions of negotiation and especially of instrumentalisation are common, and in some cases defining experiences of participatory practices – a function both of the critical accounts they offer of normative art world and research world practices, but also of the social entanglements that are core to their existence. Indeed, it has perhaps been the inherent critique if not, out-right rejection of some elements of the institution (whether this be art galleries and academics or Higher Educational institutions) that characterise these ways of working. As Suzanne Lacy famously noted in her discussion of the term 'New Genre Public Art,' its 'newness' lay in the playing out of artistic practices outside art institutions, a step from 'art in the public realm' to a 'public art.'[55]

Cecilie Sachs Olsen's accounts of the practices of her urban arts collective zURBS offers a very potent example of how these challenges play out at the intersection of creative practice and geographical research. Indeed, in much of Cecilie's writings, accounts of what the zURBS projects do become entangled with reflections on the ways the group negotiates art world and more recently, as Cecilie undertook a PhD and a postdoctoral position, university contexts.[56] She describes their practice as feeling, 'riddled with complexities relating to participant dynamics as well as institutional logics, demands and constraints.'[57] Reflecting on one commission based around experiences of local communities and housing needs, Cecilie observes the constant need to balance demands of participants and funders, and to monitor misunderstandings of intent. Whether this be participants who become aware of (and often concerned by) the contexts of funding, or equally, when they overlook them. As well as when funders assume certain kinds of activities or access that the artists' group consider antithetical to the good of the community, and their working ethos. Cecilie also details the further level of complexity added to these discussions when she moved into an academic context. She describes discovering geography as a context for her practice as hugely stimulating, as 'energizing,' and details the different ways it offered conceptual and practical benefits for the evolution of the projects.[58] As well as specific geographic ideas, this includes reconceptualisation of the idea of failure; points where zURBS projects did not work became interesting points of reflection and development; 'the research gives me a framework where I can really very easily handle failure, where it became productive.'[59] But, there were also tensions to negotiate, for the academic context placed a series of demands on her practice that were not always appropriate. Cecilie reflects, for example, on the need for research questions as a point of tension within the group: 'they said if you work from the point of a research question you already have in mind what you want and we work in order to find a question, and then the question is left open.' This prompted, as she recounted, a series of heated discussions around the work's evolution, discussions she then struggled to think through how to include in her thesis and subsequent publications. There were also practical questions, the negotiation of which prompted some serious reflection on whether academia offered what her practice needed. As she reflects, her academic role ensures that she is paid full-time for doing this work, yet other members of zURBS are not, and so the projects can become 'huge things... intensive projects, big projects on a small amount of time and money.' Whilst recognising the value of having access to funds to make work, she also observes the challenges of doing so in an academic context. Pondering the location of future projects, she posed the question;

> should you work at a cultural institution or should you work at a university? Because the thing is that cultural institutions actually understand the amount of money you need when working with artists not to because they're expensive because they aren't, but because actually they're living from it.[60]

Through Cecilie's auto-ethnographic reflections on her own creative practice with zURBS we gain a front-line account of what it means to negotiate the tensions of art practice entangled not only within society but also academic institutions. Interestingly, it seems that it is often when the coming together of creative practice and research occurs in the context of community that the most political accounts of working in these ways emerge. Sarah de Leeuw and colleagues suggest that the pace of academic progress runs the risk of us overlooking these politics. Worried by the rapid rise in the popularity of creative methods, they ask, 'how can creative and humanities based theoretical frameworks, pedagogic struggles and research methods and methodologies not fall prey to becoming tools for profit and market-drive corporate interests?'[61] They are explicitly concerned that 'without remaining vigilant and critically aware the enthusiasm for creative methods' and their rise to a 'more normative status,' might lead to them 'becoming a parody of themselves, something wholly corruptible and able to be put to use in exactly opposite ways for those for which they were intended?'[62] Focussing their attention on the interest in story-telling, they point out the very real dangers and violence enacted by the over-easy uptake of creative tools and practices within academic work. Parsing the Native story teller Lenora Keeshig-Tobias's address to the Writer's Union of Canada, they urge academics and creative practitioners to 'stop stealing Native Stories.'[63] Offering accounts of how story-telling with communities might emerge in other ways, they sketch the foundations for what seems almost a creative politics of slowness, which requires strength to resist the 'rush to embrace new and innovative means of telling stories.'[64] They explore the time needed not only to build relations, but also to build up the skills and confidence to, as they describe it, 'go unscripted' in the story-telling workings they develop.[65]

The potential incompatibility of academic cultures with those of participatory research, whether based in creative arts or otherwise, is a common point of tension. As one participatory geographer working closely with creative practitioners observed, important funding schemes might value participatory ways of working but rarely make space for their spirit and ethos.[66] Another observed, 'I was going to say I'm not good at getting grants. I mean I never applied for big grants because I don't want to say in advance what I'm gonna do. So openness is really important in my research.'[67] Other participatory geographers have reflected further on the challenges of trying to negotiate academic funding and output regimes working in the ways they do, trying to balance 'neoliberal academic pressures with our commitments to social and spatial justice.'[68] What they call for, is a finding of ways to 'negotiate relations across places and scales,' to embed a sense of 'feeling our way,' of posing other possibilities, of 'embedding site-specific struggles and frameworks within workplaces,' to enable debates about the challenging intersections of community working practices and the academy.[69] Participatory geographers writing of the UK's impact agenda asked how could participatory practices reframe the impact agenda? How might such practices offer a rescaling of values, spaces, and times of impact? Is it possible to find a way to ensure that local, in-depth, and engaged community work is valued alongside those forms of impact and to pass on research knowledge to large governmental bodies, where it ideally shapes political and economic agendas?

Heather McLean uses her combined experience as a geographer-drag artist to ensure that as academics we don't shy away from some of these difficult questions. Her sometimes irreverent and always piercing critique of academia and the art world was as evident in an interview we held as it is in her published work and in her artistic practice. Over the course of her PhD, Heather created performing drag king urban planner, Toby Sharp.[70] Toby's origin story is told as one of a response made in solidarity with artists and performers who found their work enthusiastically enrolled by feminist scholars with little awareness of the politics of what they were doing. Heather's desire to avoid becoming another 'Simone De Beaver,' a phrase she heard used to describe these scholars, combined with her own frustrations with the 'gendered performances of urban expertise that circu-lated in the neoliberal university' to form the catalyst for the birth of Toby.[71] As she elaborates, performing queer cabaret as part of academic work, including at a feminist conference, became a means 'to engage with decolonial and queer acts in an institutional space that disproportionately favours text based approaches.' Further, it offered a chance to share 'embodied critiques of gentrification and the under-acknowledgement of women's care and social reproductive labour in arts, community and education sector.' What become clear, however, as she devel-oped her project further, were the ironies of how it was received in the university world in which her research was circulating. As she writes, 'we found ourselves offering up community-engaged fare for a corporatized university that also fos-tered neoliberal knowledge economy partnerships.' [72] Even, she continues, as the performances 'satarised exclusionary community work and labour practices, we were fulfilling university knowledge mobilisation regimes that pressure scholars to extract the work of activists, residents and precarious community workers.' Eventually she began to conceive of and produce creative projects that sought to tackle this head on. Her projects evolved to include programming that offered advice to academics on how to achieve 'minority research targets' as well as to how to enrol hipster creative practitioners within your work to attract students.

Interestingly, artists and art theorists offer further means through which to consider these enrolments of art by the institution: whether this be urban planners, wider government bodies, or universities. For if public art – mainly sculptural work – was discussed as part of city planning from the 1960s on-wards, as famously critiqued by Rosalind Deutsche, then more recent forms have become enrolled within contemporary 'creativity' scripts popular in the neo-liberal city, and situated as part of programmes of urban aesthetisication associated with gentrification as well as within the wider shift in arts funding towards social concerns.[73] Claire Bishop reframes the long running discussions of art and urban aesthetics in the context of the UK's New Labour (1997–2010) government. As Bishop argues, New Labour enrolled art as part of its work on social exclusion, extending long standing agendas to increase social inclusion in the arts, by situating the arts as a tool for social inclusion.[74] As my previous re-search on arts practices under New Labour's programme Creative Britain made clear, the core question for many artists seeking funding became what can my practice do for its publics? The ideal and fundable answers generated arts projects that claimed to increase employability, minimise crime, or foster aspi-ration, often in quite neo-liberal directions. While this is often seen as an era of support for the arts, it was support of a certain form, art as a social sticking plaster focussed on social goals, rather than support for wider artistic expression and experiences. As Bishop notes, the UK was not alone in this tendency. This was an era in which all across Europe, discourses of participation, creativity, and community – once practices of subversion and engagement –emerged as the cornerstone of a post-industrial economic policy as well as, more recently, of university policy.[75]

Yet framed through questions of art's 'instramentalisation,' these debates have become more complex of late. Reflecting on her own participatory practice Dutch socially engaged artist, Jeanne van Heeswijk asks, 'why do we have to talk again about this binary position when, in my opinion, autonomy and instrumentalization are no longer oppositional strategies?'[76]According to Heeswijk, this binary presumes 'that working together with different partners such as local governments, councils, or social housing organizations invariably means that the artist is going to be instrumentalized.' Yet she suggests that in some cases this might not be true, as it might be a positive experience: 'I like being an instrument that works on self-organization, collective ownership, and new forms of sociability. I like being an instrument that enables all of us to occupy the place in which we live.'[77] This perspective is explored at length in Shannon Jackson's volume *Social Works*, a text that sits at the intersection of visual art and performance and queries the politics and practices of art's institutional and social engagements.[78] Through her case study-based exploration of a series of artworks that expose the role of 'supporting infrastructures' (from funding bodies, to patrons and arts institutions) Jackson explores how these works negotiate this role of dependence and exposure. What emerges is a set of works that tend to, Jackson argues, both foster concern with questions of 'institution,' 'system,' and 'governance,' alongside, 'telling if not fully processed attachments to concepts that could occasionally be placed as their opposite – say "flexibility," "resistance," and "agency."' Ultimately what emerges from Jackson's account is a sense, not dissimilar from Heeswijk's reflections, that interdependency might be a productive thing, there might, the suggestion is, even be liberatory potential in such interdependency, not least where revealing it becomes the focus of the work itself. These possibilities are evolved further in the next chapter which takes up the artist residency as a point of intersection with the university institution.

Coming together in community

This chapter has foregrounded the community as a site at which geographical research and creative practices come together in a range of different ways. My focus here has been the idea of participation, and in particular what happens when we think across the practices of participatory geographies and participatory arts. These are particularly interesting to think together, not least because of their shared commitment to participation (often in wildly different ways), but also because both are situated often in tension with what might be considered normative practices of geographical research and art making. Amongst the richness of these practices there are three key tensions that emerge in common, tensions that offer of value for thinking about geographical research and creative practices. First, that we need to remain open to questioning of the relationship between what happens in the process – often core to participatory working – and the nature and strategic importance of outputs for the communities being worked with as well as for artists and academics. The range of possible forms that outputs might take will be discussed further in Chapters six (on the PhD thesis) and seven (on the page). A second point that these ways of working foreground concerns critique. Both participatory forms of artwork and of geographical research have required their critics, reviewers, and assesses to reflect, at times in challenging ways, on what is important, whether that be aesthetics or more sociological aims, how can these be reconciled, and against what standards they should be judged? These are important questions that are taken further in the final three chapters of the book especially. Finally, the third point that participatory forms share and bring to the fore concerns negotiation. There are many sites of negotiation, including the university institution. This latter point is elaborated on further in the next chapter on residency: as becomes clear, whilst there are an inevitable politics from these negotiations, such negotiations can also bring about shifts in institutional research cultures, the temporalities and scales of which still require further thought.

Notes

1 Askins and Pain, "Contact Zones"; Brickell, "Participatory Video Drama"; El Khoury and Pearson, "Two Live Artists." See summary in Hawkins, "Doing Gender."
2 Tolia-Kelly, "The geographies of cultural geography."
3 See for example the cross section of examples in texts such as Bishop's *Artificial Hells*, see above.
4 See for example, Bishop, *Participation*; Bishop, *Artificial Hells*; Harvie, *Fair Play*; Jackson, *Social Works*; Lacey, *Mapping the Terrain*.
5 Bishop, "The Social Turn"; Rancière, *The Politics of Aesthetics.*
6 Bishop, *Artificial Hells*, p. 206.
7 Bishop, *Artificial Hells*; Jackson, *Social Work*; Harvie, *Fair Play.*
8 See for example, http://zurbs.org/wp/ (accessed 26/12/2019).
9 Bishop, *Artificial Hells.*
10 Ibid., p. 4.
11 Waite and Conn, "Creating a Space." See for example the special themed section of Area (2015), Practising participatory geographies: potentials, problems and politics, 47 (3). Pain and Kindon, "Participatory Geographies."
12 See note 1.
13 See discussions in Tolia-Kelly, "Fear in Paradise"; Askins and Pain, "Contact Zones"; and Hawkins, "Doing Gender."
14 Ibid.
15 Raynor, "Dramatising Austerity"; Parr, "Collaborative Film-Making."
16 Kindon et al., "Participatory Action Research."
17 Richardson, "Theatre as Safe Space"; Raynor, "Dramatising Austerity"; Pratt and Johnson, *Migration in Performance*; Kinpaisby-Hill, "Participatory Praxis."
18 I was involved in various ways in *Play Your Place* during 2012–2015. See discussion in Hawkins and Catlow, "Shaping Subjects."
19 https://www.tate.org.uk/whats-on/tate-britain/exhibition/british-folk-art.
20 Pain, "Impact."
21 Ibid.
22 Bishop, *Artificial Hells*; Sachs-Olsen, *Socially Engaged Art and the Neoliberal City.*
23 Bishop, *Artificial Hells.*
24 Brickell, "Participatory Video Drama," p. 5.
25 Brickell, "Participatory Video Drama."
26 Askins and Pain, "Contact Zone."
27 Ibid.

28 Ibid.
29 Ibid.
30 Bishop, *Artificial Hells*.
31 Tolia-Kelly, "Fear in Paradise," p. 133.
32 Bishop, *Artificial Hells*.
33 Kester, "The Device Laid Bare"; Feiss, "Response"; Bishop, "Zones of Indistinguishability."
34 Kester, "The Device Laid Bare"; Bishop, "Zones of Indistinguishability."
35 Ibid., Bishop, *Artificial Hells*.
36 Bishop, "Zones of Indistinguishability."
37 Ibid.
38 Ibid.
39 Bourriaud, *Relational Art*.
40 Brickell, "Participatory Video Drama"; Tolia-Kelly, "Fear in Paradise"; Askins and Pain, "Contact Zones."
41 Brickell, "Participatory Video Drama."
42 Askins and Pain, "Contact Zones."
43 Hawkins and Catlow, "Shaping Subjects."
44 Ibid.
45 Tolia-Kelly, "Fear in Paradise," pp. 133–134.
46 http://zurbs.org/wp/ (accessed 26/12/2019).
47 Sachs-Olsen, "Collaborative Challenges."
48 Cited in Brickell, "Participatory Video Drama."
49 Pratt and Johnson, *Migration in Performance*.
50 Jackson, *Social Work*.
51 Askins and Pain, "Contact Zones."
52 Harvie, *Fair Play*; Jackson, *Social Work*.
53 Foster, "The Artist as Ethnographer."
54 Greenberg, "Modernist Painting."
55 Lacey, *Mapping the Terrain*, p. 20.
56 Sachs-Olsen, "Collaborative Challenges"; Sachs-Olsen, *Socially Engaged Art and the Neoliberal City*; Sachs-Olsen, "Performance and Urban Space."
57 Sachs-Olsen, "Collaborative Challenges," p. 289.
58 Sachs-Olsen, Interview.
59 Ibid.
60 Ibid.
61 De Leeuw et al., "Going Unscripted."
62 Ibid.
63 Ibid.
64 Ibid.
65 Ibid.
66 Askins and Blazek, "Feeling Our Way."
67 Comment during interview
68 Askins and Blazek, "Feeling Our Way," p. 1096; Askins and Pain, "Contact Zones."
69 Askins and Blazek, "Feeling Our Way."
70 McLean, "Hos in the Garden."
71 Ibid.
72 Ibid.
73 Deutsche, *Evictions*.
74 Bishop, *Artificial Hells*, p. 12.
75 Bishop, *Artificial Hells*.
76 Heeswijk, "The Artist Will Have to Decide Whom to Service," p. 78.
77 Ibid.
78 Jackson, *Social Works*.

5

RESIDENCY

Artist Residency programmes are a particularly intensive site of intersection between artistic practice and geographical research. Indeed, artist-in-residence programmes, where diverse creative practitioners from visual artists to dancers, musicians, and writers come to work within academic institutions for a discreet period of time – from days to months and even years – are an increasingly common site for the intersection of art and research more generally. Hugely varied in location, subject, length, and the nature of the artist-institutional relations and outputs, these residencies sit within the long history of the creative and innovative ways in which artists have inhabited institutional spaces.[1] Such spaces include those of museums, galleries, laboratories, factories and, most recently universities. Indeed, since around the 1960s residencies have become an increasingly popular but also contentious means of funding art in the Western world. As discussions with a cross-section of artists who had been in residence within geography departments around the world made clear, for some they are highly desirable forms of practice, opening new horizons and often resulting in much longer term relationships, including PhD studentships, research fellowships, and publications as well as artworks. For others, however, they are a necessary evil, a means to fund the production of work, but come with the need for some difficult reflections. This includes concerns around the quality of the artwork produced, issues around uneven pay and access to resources, as well as questions of institutional 'art washing.' For the latter is not just confined to the practices of large corporations, but also increasingly found with academic institutions, for whom artists offer cultural capital, as well as valuable expressions of interdisciplinarity and forms of public engagement.[2]

Despite the growth of artists-in-residence in universities, very little critical literature exists reflecting on these practices and their politics, let alone any in-depth engagement with the relations between art and research that these residencies enable. There is however, a wealth of critical perspectives on residencies more broadly to frame this chapter's exploration of geography department-based residencies. The critical perspectives informing this chapter foreground the profoundly geographic nature of artist residencies, and help draw out the relations between research and creative practice that evolve through these university-based residencies. Three are especially relevant. The first concerns an understanding of residencies that has situated them as part of the dematerialisation of arts practices in the 1960s. In other words, the movement of artists away from the production of artefacts for exhibition, to embrace instead an appreciation that the 'work' of art emerges through the social relations that it configures, in this case the presence of the artist within the institution.[3] The second set of relevant perspectives are those around site-specific or situated artistic practice, which appreciates site as more than just a work's setting, but also the medium and material for it.[4] This enables an understanding of how some university residencies might take research practices and politics as their direct subject of study, and might, through their work, intervene within them. A third valuable set of perspectives are those which reflect on the institutional politics of residencies, how artists negotiate their role in institutional spaces and how this plays out in the politics of their work.[5] In a context in which diverse corporations and institutions, including universities, harness artists for their cultural capital, artists can feel their criticality to be compromised and their practice subsumed as part of an organisation's 'art-washing.'[6] What emerges across these three perspectives is the residency as, what we might think of after Jane Rendell, a critical spatial practice.[7] In other words, where the presence of the artist and their working practices brings (intentionally or not) a critical force with respect to the spaces within which they operate. Furthermore, as will be explored through residencies in geography departments, this critical force does not just offer a chance to pause for reflection but might also build alternative possibilities for institutional research cultures and practices.[8]

Geography has its own a minor literature on artistic residencies, whether these occur within or beyond the discipline's own spaces. Some geographers reflect on artistic groups with residency practices, such as Simon Rycroft's account of the Artist's Placement Group or Deborah Dixon's accounts of the monstrous art of SymbioticA.[9] Others offer first-hand accounts, often through edited interview transcripts of their individual artistic residencies in Geography departments and the outputs they produced.[10] Geography as a discipline has played host to a series of artists-in-residence, in departments and research centres around the world, as well as in institutions such as the UK's Royal Geographical Society with Institute of British Geographers.[11] Accounts of the residency practices and their outputs have increasingly been appearing in the pages of journals, including in the 'in practice' sections of journals such as 'cultural geographies' and 'GeoHumanities' (discussed further in Chapter 7 on the page). A further emerging stream of practice flips the script to situate geographers in residence on art projects; this includes Elizabeth Straughan in SymbioticA, and Cape Farewell.[12] These are rich and exciting literatures, and are increasingly turning attention from an analysis of the geographies of the art-works produced, to offer critical reflections on the practices and politics of the residencies themselves, however they often remain focused on a single residency and experiences of it, rather than look across residency practice. Here, I combine these perspectives with primary data from interviews and analysis of varying online records in blogs and other forms of social media to look across experiences of artists in residence in geography. The chapter begins by laying out the context for artists-in-residence in Geography departments by situating them within the wider field of artists-in-residence in universities. It will then focus on the geographies of residency, including the varied relations between research and creative practice and their relative valorisation. The chapter closes by honing in on residency as a critical spatial practice, where the space under critique becomes that of the university institution and its research practices and cultures.

Mapping the academic residency landscape

Artist residencies are diverse, multifaceted, and highly differentiated things. The rise of the artist residency in Anglo-American art world is often dated to the 1960s.[13] Key examples include the evolution of a Californian-based set of practices, including LACMA's Art+Technology Lab (1967–71, revived briefly in 2013) and Experiments and Art and Technology (1967–77), as well as the Ocean Earth Development Corporation (1980–).[14] Also crucial and increasingly recognised for its importance is the UK-based Artist Placement Group (APG; 1966–89). APG's practice of placing artists in institutions – from mining corporations, to governing bodies and hospitals – has been revisited by art historians and theorists of late, as an early example of site-specific socially engaged arts practice.[15] Common across these diverse residencies is the displacement of the site of artistic inspiration, and often production, from the studio to another space, often but not always, an institution. Focussing on the university-based residency, this discussion will begin by setting the scene for the growth of these forms of residency, detailing funding contexts, and the logics and framings that funders and institutions give to their residencies. Discussion will then evolve through a close attention to one specific residency scheme, the UK-based Leverhulme Artist-in-Residence scheme.[16] Funded between 2000 and 2016, Geographers gained over 10% of their awards, for work as diverse as sound and the city, peat bogs and fluvial geomorphology. Not only has this been a crucial scheme in developing the place of artistic practices within geographical research, but examples from this scheme constitute the core empirical material for the remainder of the chapter.

The university as residency site

The university as a site for artistic residencies seems to have emerged relatively recently, although artists have been invited to work in university contexts for a much longer time, especially, of course, within art but also with science-based departments or institutes.[17] If the US has tended to lead the way in these kinds of practices, Australian, Canadian, and UK universities have quickly followed suit, with a plethora of models from the vast university-wide scheme (see for example 'Kings Artists' at Kings College London), to smaller more departmental or research-centre focussed schemes, such as that of the School of Geography at University of Melbourne.[18] Other institutions such as MIT link together numerous high-value residency programmes, attracting well-known artists and creating content for their own campus-based arts venues.[19] It is not just academic institutions that have been funding residencies, but also wider grant-making bodies. These vary from the Leverhulme and Wellcome Trusts in the UK, to large bodies such as the US National Science Foundation which offers specific residencies within its wider programmes, including, for example, the 'Artists and Writers in the Antarctic Program.'[20] Indeed, the domain of art-science has been a key site for the funding of these residencies. The UK's WellCome trust, which focusses on supporting diverse research on health (from sleep, to wellbeing, neurodiversity as well as the latest global epidemics), is one of the first major funders to develop a programme of placing artists within science facilities. Their SciArt funding programme, from 1996–2006, funded 118 collaborative projects worth almost £3 million bringing together 'an artist and a scientist in collaboration to research, develop and produce work which explored contemporary biological and medical science.'[21] While, more recently, across Europe a model for such schemes has been offered by Swiss artists-in-labs (AIL), who supported Christina's residency discussed in Chapter 3.[22]

The conditions, aims, and framing of artist-in-residence in universities vary greatly. Some offer travel and field support, whilst others also offer a stipend – effectively a salary – for the artist. In some cases, especially within art departments in US universities, the residency does not come with pay, but offers access to university resources including studio space and making facilities, but often comes with expectations of teaching and event organisation. Other residencies offer nominal amounts – an honorarium rather than anything close to a salary – and emphasise the considerable benefits of university access, including building of a portfolio of skills and professional development as part of a research and teaching community. If some are focussed on teaching, other residencies foreground the development of public engagement with academic research, and a few focus on the artist as a researcher, exploring shared research problems with those academics they are working with. A focus on research was one of the recognised, and much celebrated features of the Leverhulme artist-in-residence scheme. From its launch in 2000 until its closure in 2016, it gained a reputation with artists and academics as a gold-standard of a residency due to both the resources it offered (including a part-time salary) and because of the vision it provided for a relationship between creative practice and academic research.[23]

The Leverhulme Trust is a charitable foundation based in London, funded through the legacy of Lord Leverhulme (who founded Lever Brothers, a lucrative soap-making business), who instructed that it should be used to support 'scholarships for the purposes of research and education.'[24] The annual artist-in-residence (hereafter AIR) scheme was one of the smallest elements of its grant portfolio, and for its sixteen years it supported between two and thirty-seven awards a year. This amounted to over 250 creative collaborations between artists, academics, and research institutions in the UK.[25] Worth between ten thousand and fifteen thousand pounds each, these grants were held for between three and twelve months by practitioners based in a huge range of academic departments and research centres. The first two awards were made to the Warburg Institute and Kew Gardens, both in London. The second round saw thirteen collaborations funded, including a poet working at the intersection of cosmology, engineering, and physics; visual artists working with neuroscientists and economists and a puppeteer working with shadow puppet theatre from Indonesia.[26] Geography secured its first award in 2004, going onto gain 32 awards for various Departments, Schools, and Research Centres/institutes around the UK (see Table 5.1). This appears to be more than any other single discipline.[27] The Leverhulme Trust wrote of the scheme in its annual reports as 'building links where the distance between disciplines is dramatic,' noting in one report during the scheme's early years that it provided a 'splendid array of interactions.'[28] Early on the intentions where clear:

> to foster a creative collaboration between the artist and the staff and/or students of that institution. It is intended that the resident artist works in an interactive way with the surroundings, and contributes to the life and work of the department or centre in which s/he is based.[29]

By 2005, the Trustees felt able to elaborate further on what they considered to be the core values of the scheme and its intersection of artist and academic community:

> The Trustees have, in making 8 awards during the year, emphasised their view that a genuine mix of contrasting disciplines or traditions between the artist and the receiving institutions is crucial to the success of the placement. The independence of the artist, ie. the freedom to follow a personal vision rather than simply to provide artistic services to the institution, is seen to be crucial to the success of the collaboration.[30]

The crucial role the Leverhulme AIR scheme played in modelling a research-orientated residency programme, and its distinctiveness in doing so, was made very clear when the demise of the scheme in 2016 was announced. The announcement was met with academic and artistic dismay. Alec Finley, internationally known artist and poet, and two-time Leverhulme AIR, praised the

TABLE 5.1 Leverhulme AIR to Geography and allied disciplines, data compiled from the Leverhulme Annual Reports

Year	Awards	Titles of projects and institutions (NB. 2004, 2012 had no awards, and 2005 had 1 but no record of what or to whom is found in the Leverhulme documentation) Titles (of individuals) and project descriptions are those given in the Leverhulme reports.
2006	3	1 Natural History Museum, Bergit Arends with visual artist Lyndall Phelps, the circulation of natural history specimens (Bergit completed a geography PhD in 2018)
		2 Royal Geographical Society, Steve Brace with artist Simeon Nelson, Early Modern Maps and the evolution of modern European Cartography
		3 Department of Geography, University of Manchester, Chris Perkins, artist Michael Mayhew, mapping and arts
2007	1	School of Geographical and Earth Sciences, University of Glasgow, Professor Chris Philo, with Artist Andrew Mulholland, musical/ sonic psychogeography
2008	1	School of Geography, University of Leeds, visual artist Jai Redman, intervening in the corporate city and environmental activism
2009	1	Department of Geography, University of Exeter (Tremough) Professor Katherine Brace, with poet Sarah Hallett, geographical intimacy, exploration and poetry
2010	1	Department of Geography (Urban Laboratory) University College London, Professor Mathew Gandy and Dr Ben Campkin with dancer Carolyn Deby, site specific urban choreography
2011	4	1 Department of Geography University College London, Dr Claire Dwyer with photographer Elizabeth Hingley, photographing suburban faith
		2 Department of Geography, University College London, Professor Matthew Gandy with sound artist and DJ Benny Nilsen, listening to the city
		3 School of Geography, University of Leeds (water@Leeds), Professor Joseph Holden with sculptor Trudi Entwistle, exploring water use and engagement
		4 Marine Institute, University of Plymouth, with writer Philip Hoare, writing the sea
2013	4	1 Department of Geography (Urban Lab) University College London, Dr Ben Campkin, with photographer Rab Harling, architecture and communities of the Belfron Tower
		2 Countryside and Community Research Group, University of Gloucestershire, Dr Owain Jones with artist Mr Antony Lyons, watery landscapes
		3 School of Geography, Politics and Sociology, Newcastle University, Dr Alison Williams with visual artist Dr Matthew Flintham, geopolitics of air space
		4 Department of Geography, Durham University Dr Jeff Warburton with artist Laura Harrington, fluvial geomorphology and peat bogs

2014 6

1 School of Geography, Earth and Environmental Sciences, University of Plymouth, Dr John Martin with designer Dominica Williamson, visualising information about natural landscapes

2 Department of Geography (Urban Lab), University College London, Dr Andrew Harris, with photographer/ multimedia artist Tom Wolseley, visions of vertical London

3 Department of Geography, University College London, Dr Tariq Jazeel with writer Amita Murray, text and place

4 Department of Geography, University of Cambridge, Dr Bhaskar Vira with photographer Toby Smith, water crisis and the Indian Himalayas

5 Department of Geography, Durham University, Professor John Wainwright with Miguel Santos, transdisciplinary artist and researcher, rivers

6 Department of Geography, Birkbeck, University of London, Dr Karen Wells with Ain Bailey, sound artist and DJ, sound art and public space

2015 4

1 School of Geography, Queen Mary, University of London, Professor Alison Blunt with installation artist Janetka Platun, home and migration in East London

2 School of Geographical and Earth Sciences, University of Glasgow, Professor Maggie Cusack with visual artist Rachel Duckhouse, nano-architecture of nacre (mother of pearl)

3 Department of Geography, Royal Holloway, University of London, Dr Harriet Hawkins with visual artist Flora Parrott, subterranean spaces in human and physical geography

4 Department of Geography, Durham University, Professor Rachel Pain with songwriter and musician Brenda Heslop, song writing and housing

2016 5

1 Department of Geography University College London, Dr Claire Dwyer with theatre practitioner Tom Bailey, creative approaches to refugee research

2 Department of Natural Sciences, Geography and the Environment, Manchester Metropolitan University, Dr Michael Gallagher with sound artist Dr Mark Wright, sound walks and biosensory media

3 Department of Geography, University of Sheffield, Professor Andrew Hodson with multi-media artist Naomi Hart, ice and global warming

4 The Geography Department, Swansea University, Dr Amanda Rogers with theatre maker Dr Bridget Keehan, soundscapes and place

5 School of Geography, Politics and Sociology, Newcastle University, Professor Rachel Woodward, with movement practitioner Paula Turner movement and research

scheme in Artist Network's a-n magazine, saying, 'it is possibly the only funding which pays fees for artists to develop research in new contexts without expecting the results to be defined in advance.'[31] A Scottish Arts organisation linked this research ethos with a commitment to resourcing artists:

> the scheme is also important because it paid artists directly and at a rate that was commensurate with the investment of time in research. The importance of this point cannot be underestimated. No other schemes that involve artists with academics actually provide sensible funding for the artists.[32]

For academics the scheme represented a welcome antidote to the increasing tendency to use artists for impact and the fetish of interdisciplinarity which rarely resulted in adequate resourcing. One biomedical researcher noted online that the Leverhulme AIR was 'one of the only schemes that faced interdisciplinarity head-on and accepted that evidence of mutual interest and respect were the relevant criteria for awards.'[33] Others echoed this, suggesting that this was one of the few schemes that responded to the trend for interdisciplinary research by really 'put[ing] its money where its mouth is.'[34] In its own very brief discussion of the scheme's closure, the Trust cited issues around funding pressures and the relative success rates of other schemes.[35] For the art world this was a further blow, indicating the lack of value and support for these ways of working, and evidence of an intensifying land-grab from other disciplines for whom the arts offer a public engagement service, rather than a form of research.[36] The online reports demonstrate that the scheme was worth around 0.5% of the Trust's annual grant-making budget, a fact not lost in the community, one artist took to Twitter to note 'its probably a scheme that invested between £250k-£500k out of a total annual grant making of something like £10 million annually.'[37]

The closure of the scheme was a loss for geographers and creative practitioners seeking to work with them. Geography Departments/Schools and Research Centre across the UK were major recipients of the Leverhulme AIR.[38] Across these awards many different institutions were funded, with some getting more than one award: UCL Geography and their Urban Lab secured eight grants; while Geography departments or schools in Glasgow (early adopters) and Newcastle each got three; and Exeter, Leeds, and Durham each received two. The scheme funded, as Table 5.1 demonstrates, a rich range of creative practitioners, from sound art to theatre, dance, visual art, songwriters, and musicians as well as photographers, filmmakers, and poets. The Geographers involved too were equally varied, from physical geographers interested in glaciology, crystallography, hydrology, and soils, to human geographers working on biogeographies, suburban faith landscapes, urban soundscapes, and underground imaginations. Many, but not all, projects have a substantial online presence, mostly through the practitioner's home pages and departmental news pages, offering an archive of records of grant awards and launches as well as adverts for events and exhibitions. More recently, residencies have active blogs, Instagram and Twitter accounts that make visible elements of the residency as it happens.[39] In addition to this documentation, there are a host of artist objects and artefacts produced, as well as exhibitions and events staged. These residencies also often produced book chapters or papers detailing the residency and its 'results.' Sometimes these publications were experimental in form – visual essays, for example – other times taking the form of a conversation or interview, often taking advantage of the growth of 'creative' and 'in practice' sections of journals (see Chapter 7 on the page for further discussion of these sections).[40] In many cases the residencies came to shape continued conversations between the creative practitioner and geographers, finding form in ongoing collaborations, whether this be practice-based PhDs, an array of small and large grants, or ongoing event series and more informal collaborations.[41] Given the importance of the residency as an intersection of creative practices and institutionalised academic geography, not least during the residency but also in the years that follow, I want to take a closer look at these residencies and some of their dynamics. What comes into focus is the rich range of relations between research and creative practice that these residencies encompass.

FIGURE 5.1 Installing the void at the Association of American Geography Conference, Boston, 2017. (Image, JOl Thompson)

Arts and research relations

While the closure of the Leverhulme AIR scheme was a blow for geographers and creative practitioners alike, the ensuing cross-disciplinary debate brought to the fore the value placed on its modelling of art as a research practice, and its appropriate pay. While it has been held up as a gold standard by artists and academics alike, there is a rich range of different relations between art and research within the Leverhulme AIR as well as across other residency programmes. To explore these issues, I want to begin from an experience I had during an exciting event led by social scientists and scientists exploring climate change and the arts and humanities. One of the key talks, early on in the event's two days, given by a very influential supporter of arts and humanities work within climate change, offered up a certain imaginary of the place of arts within this field that shaped the tenor of the rest of the event. This imaginary is best summed up through describing an image that was projected on the screen for much of the talk. The image was of an arrow – which represented the progress of a research project – moving from (on the left) activities such as hypothesis forming, data collection, and analysis to (on the right) dissemination and engagement. Arts was situated very firmly on the far right-hand side of this arrow, downstream from knowledge making and part of knowledge dissemination and public engagement. A later group discussion explored the book, *The Myth Gap*, which makes a call for better and more myths – stories – as a means to engage publics with the challenge of climate change.[42] This was seen as a very positive thing, after all dissemination and engagement was what would change minds, and the image was shared in good faith and received enthusiastically, there was a real desire to make space for the arts within climate change research.

This particular understanding of the role of the creative practitioner appears to have become dominant within a series of institutionally funded residency programmes. If we look for example, at the information offered on the Department of Geography, University of Melbourne's website about their residency programme (AiRG), we can see an explicit focus on 'building links between research within the geography department and the wider publics.'[43] The programme asks for a six months residency, for which there is a $5000 AUS stipend, and an expectation of several seminars and other engagements with researchers, and a heavy focus on public impact and outreach. From the outset the application notes focus on how the programme 'aims to generate new and stimulating modes of communication about the work done in Geography and generate discussion amongst broader audiences.'[44] This is reinforced in the notes for applicants, where it states, 'because the development of effective communication strategies to engage audiences is central to the AiRG, the artist will also, during the residency, participate in the launch of AiRG event; Blog and/or Instagram during residency to communicate progress/ process; propose and general processes that facilitate the publication of the artist work for non-experts.'[45] One of the core goals was,

> that artist and staff member/s will work collaboratively to explore the complexities of the identified issues through a geographical perspective.

Together they will discuss the modes of artistic expression appropriate for communicating and generating discussion about these issues to broader audiences.[46]

This sense of engaging 'wider audiences' is often a key justification for University's AIR programmes. Kings, London's residency programme 'Kings Artists,' announced in 2017, to support four artists a year with five thousand pounds, offers an interesting demonstration of the complex messages that emerge around what residencies are and do.[47] The residency programme stems from the university's 'Arts and Culture at Kings Strategy.' Here, an institution without any creative practice departments commits itself, in a text steeped in the language of Higher Education policy priorities, to supporting the role of arts and culture as 'offering distinctive opportunities to students and academics, helping to deliver world-class education and research that drives innovation, creates impact and engages beyond the university.'[48] In talking specifically about the artistic residency element of their work, they highlight the twenty-eight residencies that have been supported over the last fifteen years in a range of departments and with a range of 'outputs,' textiles, video, 3D printing, etc. For prospective applicants they detail the breadth of possibilities, noting,

> residences provide opportunities for artists to develop their practice within an academic environment, in collaboration with researchers, and provide valuable learning models for both staff and students. They support cultural and artistic exchange, nurture experimental and new ideas, support research and help develop new work and approaches.[49]

A closer look at their 'Kings Artist's tool-kit,' designed to 'help ensure the smooth passage of the artist as resident within the university,' finds the artist emerging as a figure of 'connection.' For example, 'Residencies can play a powerful role in making connections across disciplines, between the university, the people and the organisations around it, between different communities and across different ways of thinking.'[50] Much of the guide focusses on legal agreements, contracts, and start-up memos. Only once in the guide is the concept of practice-based research suggested and despite head-line descriptions, the overwhelming takeaway message is one of residencies as tools for public engagement and teaching development. This is clear under the guidance on expected outputs where they detail things the residences should achieve, including 'community engagement projects... an output such as a book or app... an event such as an exhibition or theatre production.'[51] Setting aside the concerning issue around whether five thousand pounds is anywhere enough money to cover the time and costs of achieving an exhibition or production, the message is that these are residencies aimed at public engagement and impact and the evolution of teaching.

The intended focus here is clear; the artist is, to borrow the Leverhulme Trustee's language, a 'service provider' for academics and their institutions. There is, on the face of it, nothing wrong with this, after all many creative practitioners, as the previous chapter demonstrated, make work that is very much focussed on engaging with publics and stakeholders in a rich range of ways. If however engaging with publics is the only part of a research process into which artists are admitted, this is somewhat limiting. Furthermore, geographers have, for a while now, been sounding bells of caution when it comes to thinking through the location of artists as part of contemporary university agendas around impact and public engagement. Indeed, writing of the rise of arts and visual culture practices within Geography in her Progress in Human Geography report of 2012, Divya Tolia-Kelly observes that geography must guard against seeing 'art' as an easy component of the "impact agenda."[52] She continues, 'in an era where public impact and engagement are politically encouraged, there is a risk of collapsing the differences between visual culture as a discipline and the visual as an accessible mode of research communication.'[53] More recently, these concerns are echoed in worries around the 'art-washing' of universities, with higher education institutions following city planners and corporate entities in harnessing the cultural capital of often poorly paid artists to build their own image. [54] The written work and collaborative performance practices of Geographer and drag artist Heather McLean, as discussed in the previous chapter, offer critically astute commentaries on how universities and individual academics have been participating in drawing down the cultural capital of arts practice, as part of branding and promotion exercises and to tick boxes around impact and interdisciplinarity.[55] McLean's account observes how oftentimes universities and individuals will enrol precarious artists into their research projects and events, extracting labour from the creative sector, often at far cheaper rates than they would pay a post-doctoral researcher for example. This 'outsourcing' is, as scholars and artists point out, often going on whilst these same institutions are 'defunding their own arts activities and stretching arts faculty to breaking point such that their own arts practices are no longer sustainable.'[56] Individual academics often don't help, failing to value and respect the place of practitioners as researchers in their own right. Geography is littered with examples of presentations and visual essays which fail to properly acknowledge the artists who produced their images at the core, and whose academic-authors are all too ready to harness the products of artists to enhance their own reputation as part of academia's growing 'star' culture and use of social media as a self-publicity machine. There is a risk, it seems that only certain types of arts practices become desirable within the university, those that help promote the academic work of other more fundable scholars, all the while devaluing the arts activities of faculty members, and the arts and humanities skills of others whose research is not so easily packaged up into the 'trends' of current funding regimes.[57] What follows probes a series of examples of artists in residence to explore some of these intersections of research and art in more depth. The aim is less to valorise one form of intersection over another, but is instead to direct attention to the ways that, in fact, in many cases the residency practices refuse a singular position upon any imagined arrow of research directionality and intentionality.

In the case of the Leverhulme AIR scheme, what was set up as the "ideal" offered by that scheme by the Trustees, and by those I spoke to, was its situation of artists very firmly as part of the process of doing research, often a coming together around shared research topics. On hearing that the scheme was to be closed, an artist tweeted, this was 'one of the few on-going and established schemes that supports artists to work with other researchers across the natural and social sciences as well as the humanities.'[58] Indeed, as this commentator goes onto note

> it is one of the few opportunities which really understood that artists collaborate within other research discussions, and should start not from an assumption of illustration and public communication, but from first principals of mutual interest. It provided time for genuine dialogue, and for unexpected results.[59]

This echoes many of the accounts of residency that I heard. For many, the artist was conceived of as researcher from the very start, with the residency about making space for research discussions between individuals with very different practices and perspectives. There was much emphasis on the role of a shared interest in common topic or place. In the case of Matthew Flintham's residency, hosted by Alison Williams in the School of Geography, Politics and Sociology at Newcastle, there was a shared interest in militarised air space.[60] Both had previously explored questions of how we know airspace and were interested in how we can map the overlap of different aerial territories, as well as wider questions around how we know the invisible – Alison, from her perspective, as a geopolitical scholar, whilst Matthew, from his, as a sculptor and visual artist. In interviews both of them were clear about the 'fit' between their shared interest in the topic, and the excitement they felt at the different resources the other would bring to the discussions.[61] Interestingly, both had some previous experience of valuing the other's ways of knowing: Alison had worked with artists previously on small commissions, while Matthew had done an Arts and Humanities Research Council PhD studentship with film-maker Patrick Keiler and Geographer Doreen Massey.[62] Exploring the project with the two of them, it was clear that the production of research in common, was not seen by either as antithetical to an approach that also created a site for wider academic and public engagement with their work. So as well as a book chapter, they also produced an exhibition that had explicit goals of public engagement (discussed further in Chapter 8).

In the case of other residencies, it was a commonality of research methods and approaches that brought artists and geographers together. Participatory Geographer, Rachel Pain, reflected on a Leverhulme Residency that brought together folk singer-song-writers from the group Ribbon Road, as well as a local photographer Carl Joyce in the School of Geography at the University of Durham.[63] Here creative practitioners, alongside third-year undergraduate students, were part of a participatory research project exploring contentious local housing issues in Horden Colliery, County Durham, North East of England.[64] Reflecting on these practices, Rachel observes,

> that tradition of folk is all about listening to people, but it is also about music for people, for working people. So there's alot of parallels with participatory action research as well... and of course it was just lovely and beautiful and incredible in the way it captured the situation in the village.[65]

It was this shared sense of methods and approach that led Rachel to continue to work with Brenda from Ribbon Road on a later project on trauma and domestic abuse. As Rachel observed of the participatory group practice they developed with domestic violence victims,

> something like trauma, nobody understands it, so how do we raise aware-ness of it, so the songs are like an output, but what we've ended up doing is Brenda's plays some of her oldest songs... its just been incredible, the way the music opens being up, its really powerful. She thinks carefully about which song she thinks the group is ready for, that just kind of get everybody's emotional juices flowing... a lot of the literatures are saying you know in different ways trauma, it is kind of inexpressible... it is often spoken about on behalf of the people who have been traumatised... and yet... this really helps out.[66]

These observations direct us towards the complex relationship between the songs and their role in the research process. Brenda playing her existing songs creates the atmosphere for the difficult things the participatory research group are going to be thinking about together – 'music opens people up.' This is an important part of the research process enabling people to get in touch with something of-ten thought of as hard to get at. The collective song writing also sits alongside drawing and creative writing as a means by which the group explored some of the detail of their experiences. Working with song writing was seen to enable experiences of trauma to be explored and represented 'with' those victims, rather than 'on behalf' of the victims. As Rachel continues,

> she just wants to help represent their stories.... And she just comes up with this beautiful, beautiful work, that kind of does that, and they think it's fabulous... they're researchers asking questions, trying to produce knowledge.[67]

As well as thoroughly embedded within the research process, Rachel's and Brenda's accounts of the project also emphasise the force and value of outputs. Ribbon Road's website includes, for example, the following news item:

> The New Album will be released on 28th September. It contains the re-sults of work done during 2018 with Rachel Pain of Newcastle University,

looking into trauma associated with domestic abuse. Brenda and Rachel worked with a group of women in My Sisters Place in Middlesbrough.

The songs were inspired by the sessions, the stories and the feelings which emerged from the brave women who now like to be known as the Paper Dolls group. *Paper Dolls, holding hands, Hang together, try to stand Paper Dolls, do not cry Hang together, stay alive* [my italics – these are lyrics].

> Paper Dolls contains 5 newly written songs, 2 un-released songs, and new versions of 2 songs which originally appeared on the CD 'It Couldn't Last' which were used to inspire the sessions.[68]

As well emphasising the importance of these songs for the research process, Rachel also indicated the role of this process for producing material that formed a valuable part of research reports. This included a nationally disseminated report for charity that was also presented at the Women's Aid National Conference. As Pain reflects, 'the song-writing will get interest, you know, touch people, so they feel it before they start reading it, which I think will make a difference.'[69]

It is clear from accounts offered by those geographers and practitioners involved in the residencies, that creative practices can be distributed throughout a research process, not just confined to one end – the dissemination – end of it. Across geography there are examples of creative practitioners in residency as both part of research processes that have evolved from a shared research topic, as well as offering valuable public and stakeholder outputs. Clearly there are myriad benefits of artist residencies, and the art and research relations within them are equally diverse, and often evolve, as artists and geographers come to work in proximity. Yet it is also clear in other cases that residencies can go awry, perhaps were expectations are not clear, or don't align, and indeed there are examples within and beyond geography where this has clearly happened. To close, I want to explore how these residencies themselves became sites at which to explore the spaces and cultures of University research that we work within. How, in short, might these residencies offer a site of institutional critique, which might consciously or unconsciously come to reshape our institutional spaces and practices?

Residency as critical spatial practice

The recent reassessment of the activities of the Artist's Placement Group in the UK in the 1960s has situated them as an early example of socially engaged practices, an example of site-specific research-based working and, for geographers, as an example of how artistic practices might 'do impact.'[70] Founded by Barbara Steveni and John Latham, APG called itself a range of things: an 'artist consultancy,' a 'network consultancy,' or a 'research organisation' arranging 'placements' for artists within both public and private organisations for limited contract periods.[71] These organisations included the British Steel Corporation, Ocean Fleets Shipping Company, the UK Government Department of the Environment as well as the Department of Health.[72] The results of the projects ranged from art education, on-site installations, public outreach, and creative uses of technology in the context of critical reflect on management, working practices and policy.[73] Their effects have been far reaching, as Bishop writes,

> what needs to be appreciated today is APG's determination to provide a new post-studio framework for artistic production, to create opportunities for long-term, in-depth interdisciplinary research, to rethink the function of the exhibition, and to create an evaluative framework for both art and research that displaces any bureaucratic focus on immediate and tangible outcomes.[74]

It is easy to dismiss, from the perspective of the contemporary popularity of socially engaged participatory and site-specific practices, the importance of how APG conceived of the artist as having a social function. In place of the artist as producer of material artefacts, which was still the predominate sensibility in art of the era, APG foregrounded the intention of the artist to engage with their 'social field.'[75] Coining the description, 'incidental person' they sought to shift attention from the making of objects to the creating of incidents or situations through the presence of the artist.[76] Art historian Howard Slater frames this as less about the production of 'a static subjectivity where the artist's person, commodified, becomes an institutional currency,' but rather about their performative conceptual activity, 'in becoming immersed in the unfurling dynamics of the workplace.'[77] Instead of creating 'contemplative still lives' of the institutions in which they were situated, they were encouraged to maintain a position of independence that would 'release the impulse to act' to potentially shift the conditions they found in the institution.[78] Based on accounts of experiences developed during the three different Leverhulme AIRs, as well as analysis of the on- and off-line documentation of these residencies, I explore how the AIR could be considered incidental people, offering sites of reflection and even intervention into the everyday practices of geographical research, departmental research cultures, and the cultures and practices of academic conferences.

Everyday geographies

> Being in a very large institution can be intriguing for one used to practicing mainly in community halls, hospital wards, day centres, classrooms, gardens and fields. I have clocked up miles wandering around and through and over and under the "territory" that I now call GPS (School of Geography, Politics and Sociology, keep up!) [entry from residency blog].[79]

Paula Turner, a dance and movement practitioner resident in the University of Newcastle's School of Geography, Politics and Sociology hosted by Rachel Woodward, was interested in researching movement within geographical spaces and practices. She suggests, following the words of Simone Forti (An American Italian Post-Modern Dance artist) that she was interested in how through her practice she could 'get research to move.'[80] While the languages and practices of Paula's work and her blog feel familiar to those cultural geographers working on embodiment and movement, including thinking through the place-ballets (after David Seamon) of geographical fieldwork, her residency applied these less to dance spaces or events, than within the everyday spaces of geographical research.[81] While these spaces included those of the lab, the field, they also included teaching spaces. In one case an undergraduate seminar on post-structuralism was transformed from being based on a reading about Foucault to using intergenerational dance practices as a means to explore experiences of confinement, including those of elderly people.[82] In another case, Paula and her group worked with staff and students to create body mappings of lecture theatre spaces, encouraging reflections on the form, function, and 'feeling' of the 'learning environment.'[83] More often though Paula and the intergenerational dance group the *Grand Gestures Dance Collective* occupied corridors, hallways, and stairwells.[84] These are not uncommon spaces for artists in residence in Geography departments to find themselves exploring: the 'a's, b's and c's' (attics, basements, and cupboards) as Kate Foster and Hayden Lorimer put it.[85] For Foster and Lorimer the attraction of these spaces was that of free-er operation; 'The guardians of these valuable local repositories and unintended archives (sometimes rendered 'uneconomic') were most important allies. With trust fostered, free rein to use overlooked and generally outmoded resources was possible.'[86] Discussing Paula's choices with here, it was not just expediency that draw her here, far from it, in fact these spaces turned out to be quite difficult spaces to work in. Working with her dance group they mounted an 'expedition' to the stairwell, setting up camp on a landing and stopping people to talk to them as they passed, they were also drawn to the large table in the building's entryway.[87] This was a rendering unfamiliar of these spaces, disrupting of the normal 'flow' of traffic to explore these 'curious place(s) with all of their comings and goings, automated talking doors, lifts, openings, closings, hellos and goodbyes.'[88] A video included on Paula's website records one of the pieces made during these investigations. Entitled *Prime: the first of five scores for movement*, it was described as 'a selected fragment from a durational loitering in a threshold space.'[89] A large silver bowl of water, a cloth, two dancers, the table, and many passers-by, interact to form a four minute forty-seven second story of this space, a daily part of the research life of the school but a part not much remarked upon. As Paula notes 'previous observations led us toward considering how the space was tended and maintained and how a table is still an invitation to commune, even silently.' Yet while these spaces were often overlooked, and despite the fact they were clearly supposed to facilitate work and communication between people in the university, it became clear that only a certain type of 'work' was really sanctioned in these spaces.[90] As Paula detailed in an interview, these practices very quickly became a site of conflict when the specific movements of their group were deemed 'out of place,' and 'what are you doing?' became a refrain from a range of staff in the building until finally security were called and they were asked to leave. As well as exposing what was seen as 'appropriate' work in these spaces, what also emerged was the securitised and monitored nature of these locations.

During her time in the department Paula was not only interested in how she could put the place to work to 'get research to move,' she was also interested in drawing out the labour of other practitioners in the building. Attending to those who tend and maintain the building for the researchers became part of how she sought to understand the building and the 'remarkable tangible effects of work that quietly maintains people and places.'[91] In her project, *Quotidan mapping- the world in bucket* Turner conducted a 'mapping with Pauline,' the cleaner of the Daysh building in which the School of Geography is based.[92] She is interested in the 'ticking along, the ever present work of maintenance and care that Pauline puts into the fabric of the building.' Conducting an ethnography of Pauline's work over several weeks, Paula built an account in video and words that foregrounds the 'ease and competence of her movements' her social connections 'she seems to know everyone,' her 'embodied knowledge of the territory gives an informality and a welcome to what is quite a formal straight laced kind of building where social interaction is not always easy.'[93]

These were ongoing, often low-key interventions, often developed on the spot in these diverse locations within the university. Yet the reactions they prompted, including the range of questions that emerged suggests the residency's force in questioning the codes and practices of the spaces we work in, and the different forms of labour that go into maintaining them. Paula's presence as an incidental person situated the spaces of corridor, stairwell, and cleaning cupboard alongside the lab, field and seminar room as part of the production of geographical research. As well as enabling an attention to the spatial practices of department research, residencies also intervened in departmental research cultures in ways both expected and unexpected. It is to these interventions I now turn.

Cultures of research

The presence of the artist in residence within geography departments in many cases has effected how scholars within these environments understand what research can be, how creative practices might sit within research cultures, and has expanded the sorts of collaborative interactions many people have. As with wider theories of artistic change, this is not necessarily to suggest huge dramatic changes, but is to note smaller, sometimes more ephemeral shifts in understanding and thinking about what geographical research might be considered to be and the kinds of practices geographers embrace.[94]

During her Leverhulme funded residency at University College London's Geography Department, writer Amita Murray, organised a series of writing workshops. Host Tariq Jazeel describes a core group of about ten individuals who regularly attended fortnightly creative writing workshops, mostly novices, who also shared their writing on a blog and at the end of a year of workshops produced a book.[95] When asked about the effects of the project he described it in terms of 'many flowers that bloom.' He could detail specific outcomes, the book, as well as vignettes some wrote in the sessions that made their way into academic papers, a research associate who had been starting to write, but ended up so inspired she won a writing competition as an 'emerging voice.' When pressed further as to his own experiences he reflected on two things. The first was a sense of the shifting of the departmental cultures of expertise: where post-docs and staff were all in it together. He also reflected at length on the wider shifts these workshops helped instigate in discussions about research culture:

> it became a space where we talk about the presence of writing. So you open up a set of conversations about writing. And I've thought that, that geography is a peculiar discipline in the sense that we strive towards the humanities, the social sciences and the sciences. But I think actually that I think we were much more social sciencey than we would like to think even when we dabble in the humanities. And so we think about writing as writing up. We don't think about writing as, as necessarily key to, to what it is we're trying to do in the arguments we're trying to make. So we don't think, in other words about the kind of the form, what it is that we do, but we don't have those conversations often enough. And I think that that was really useful. This residency was really useful to open up a set of conversations about and the importance of narrative, the importance of form in building the arguments we're trying to make rather than just that sort of model of you go out and do research in the world, come back into the office and you write that research up, which is very sciency. it's a model of writing that comes from the scientific method I think.[96]

Similar discussions about geographical knowledge were raised in the context of my experiences of having Flora as an artist-in-residence within the Department of Geography at Royal Holloway, University of London. The department has a history of artists being present, including as research associates on funded research projects, but also through programmes such as the *Landing* exchanges.[97] These exchanges curated by Ingrid Swenson, paired artists and geographers in conversation, and resulted in the exhibition and book *Landing* that not only presented the works produced but also included conversations on practices of collaboration.[98] With a cross section of human and physical geographers used to the presence of artists, Flora's fieldwork, the drawing workshops she ran, her presence in the lab, her invitations to be part of events and installations were welcomed. Indeed, since her arrival in the department, the intersection of physical geographers with artists has become richer, resulting in a series of small grants, journal articles, exhibitions, and a growing inclusion of artists as part of large grants to the science funding councils. It was not just within geography departments however, that these residencies might be thought of as shifting departmental cultures, but also beyond, in wider disciplinary research cultures. To close this discussion I want to consider a 'conference paper' Flora and I gave at the Association of American Geographers conference in Boston in 2016. I use the phrase conference paper with caution here, as the discussion will demonstrate.

Installing the void

We inspected the room, smoothed our hands over the bland hotel wall-paper, sighed with dismay at the flatness of the walls and the light fittings. We tried to use section cups to suspend the canopy from the walls, they would not stay attached. We piled up chairs, they fell over. We eventually enrolled friends to hold up the canopy for us. When the time came for our 'presentation', we passed the fabric over the heads of the audience, spread out more fabric on the floor. Projected some instructions in place of a standard power-point, 'when the mirror reaches you, look through your phone into it.' We wanted people to see themselves in the midst of the darkness created by the canopy. I sat cross-legged on the floor and began to read, a text composed of found materials combined with reflections on fieldwork conducted in caves. A live video of me sitting, apparently at the bottom of a hole, reading the text was projected onto the screen in front of the audience. The mirror never got passed around. People seemed bemused (Figure 5.1).

From field-notes reflecting on conference experience, April 2017, Boston

Our conference-paper-cum-art-installation was an attempt to install in a conference-hotel-seminar-room a series of voids, using fabric, mirrors, projections, and texts. It was just one of the many activities Flora and I developed during her Leverhulme residency about the underground. Whilst we had written papers, produced exhibitions, curated workshops and a range of events, these had mainly, we realised at some point in our discussions, sat easily within either an academic context or within an art world context. We wanted to push ourselves, to find something that felt more like a meeting of the two of our skill sets. The result was what became known as the 'voids' paper. This was as much about exploring voids as it was about conducting an experiment with pushing at the form of 'conference paper,' at seeing how far we could go in response to those increasingly common calls for 'other formats of presentation.' Almost every major conference within geography, but also within sociology, archaeology, and so on, now includes a series of calls for 'papers' that suggest 'alternative presentation formats welcome,' encouraging the use of video, sound, and other forms in the conference space.

In a sense then what we were doing was not so new. Yet, more often than not, whilst fascinating, we found that oftentimes creative practices appearances in geographical conferences could feel heavy on the explanation and felt a bit like creative 'show and tell' – 'here is this creative practice I produced at another venue'. Flora and I wanted to do something a bit different; in short we wanted to create a multi-sensory art installation in that conference. We used our fifteen minutes perhaps oddly. We had no slides, no explanation, no real introduction or conclusion. We used our time to install our void, attempting through simple

materials to create a different visual, sonic and affective environment in the conference space. I was concerned at the time, and still think several years later, that
the effect was mainly of a bemused interest, polite attentiveness, or outright confusion, not helped by our slot as the final 'talk' at the end of a day in one of the
airless boxes that these conferences occupy. I was used to presenting alternative
forms of work, but the experience of rendering a conference room as the site
for an installation was different. It made very clear that despite calls for alter-
natives, these were very much audiences used to certain kinds of languages and
logics; you face the front, you listen, maybe you laugh, sometimes you look at a
power-point. As such, as spaces that move from interested engagement to fidgety
jet-lagged boredom, these conferences are challenging environments in which
to shift the practice of conferences goers from a sort of didactic listening mode,
where we expect to be told about research and findings and arguments, to one of
sensory experience and investigation, in which we ask people to think-feel our
installation-as-paper.

We were not the only ones to take up the organiser's request for alternative formats; Sarah De Leeuw, a geographer-poet, had responded to the session's provocation with an extended poetry reading for her fifteen minutes; Anna
Secor and Jess Linz had presented a psychoanalytic skit as their means of exploring void-ness. As such our installation was in interesting company, it felt when
we reflected on the experience that perhaps the audience found experimental
papers that were framed through words easier to engage with than paper that
required other kinds of engagement too. The session organisers were generous
enough however to welcome our contribution with open arms, and when it
came to the follow-up publication were happy to work out how to accommodate
the translation of the installation of the void onto the page. Following experiments with layout and Flora's guidance, our chapter starts and ends with full page
images (see Figure 5.2) before offering an edited version of the spoken-word text

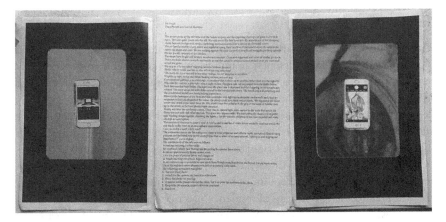

FIGURE 5.2 Writing the void: draft pages from the book chapter. (Image, author's own)

of six voids which had filled much of our fifteen minutes. We struggled to work out how to produce an abstract and key words that continued the experimental form, and ended up resorting to a point of explanation: 'in this chapter we re-call our attempt to install six voids in the Association of American Geographers conference.'[99]

Conferences are often understood to have four key functions as academic knowledge spaces; they are part of intellectual communication, professional so-cialisation, the reproduction of academic hierarchies, and the legitimation of new subfields or paradigms.[100] Conferences have been explored as sites of ac-ademic knowledge claims; 'conference participants learn from the conference programme what kinds of knowledge and means of claiming such knowledge (are) valued by the academic profession.'[101] If conferences are sites where we can test out, expand and develop, but also help to normalise forms of academic practice, then our experimental installation was an important moment for us, in terms of trying to push at both what is possible to do with a conference paper, but also in terms of realising what these kinds of practice-based research requires from geographical audiences. As such, it was clear from the audience reaction that this was perhaps not a presentational form that people were used to engaging with. Our guidance, projected onto the screen at the start, about how to interact was ignored, we pondered whether we are asking a lot, for the audience to grap-ple with an installation as part of the content of the work, not just a background setting in which to give the paper, or something the paper talked about, but, in effect the paper itself.

We are sure some people did not grasp what we were doing, others were openly puzzled and baffled, there were positive comments on the text, but little about the immersive environment we had created. It felt a bit like we had been asking our audience to do something many of them likely do very well in gallery settings but they perhaps just don't expect to have to do in a conference session, plus, it was, afterall, the last paper in the last session of the day.

Residency

This chapter has taken the artistic residency within geography departments as a focus of attention. The residency is an important part of the contemporary fund-ing landscape of creative practice, and as is clear there is a huge cross section of schemes that have engaged creative practitioners as residents within geography departments as well as wider university contexts. For many of those developing and participating in such residencies, these residency programmes intersect in different ways with the tense origin stories of practice-based research as a func-tion of the restitution of artistic research into academic contexts. Indeed, it seems residency logics offer a microcosm of the variety of interactions of research and creative practice. In discussions of residency practices, and especially in the wake of the closure of the UK Leverhulme Trust Artist-in-Residency scheme a series of 'ideal' relationships between research and creative practice seemed to rise to the fore, especially that which foregrounded resident as researcher, rather than resident as a service provider. Yet what the accounts of such schemes and experiences of them suggest is that often within a residency might enfold a number of different forms of research and creative practice relations. Further, as the final section turned to consider, the presence of the resident in the department might offer a valuable lens through which these relations can be discussed in more detail. One of the ongoing areas of debate across arts and humanities and geography is how we talk about the 'work' art does, how we understand the means through which the world-making and shaping effects that are claimed for these practices not only come about, but for how long they persist, and further more at what scale do they become significant.[102] This is not something that has preoccupied this discussion or this volume more broadly; thus, it is beyond my scope here to claim any in-depth shifts in research culture as a result of residency programmes, but it is clear that residencies offered for many of those involved an important point of thought around what these research cultures are, and what relationships between practice and research could be. Perhaps the clearest evidence for this is offered by the amount of residents who go on to have a longer-term relationship with the departments in institutions in which they are based, often through further research on grants or through practice-based PhDs – it is to such PhDs that the following chapter turns.

Notes

1 See for example, Grzelec and Prata, "Artists in Organisations"; Stephens, "Artists in Residence"; Harris, *Art and Innovation*.
2 Bastashevski, "The Perfect Con."
3 Bishop, *Artificial Hells*; Lithgow and Wall, "Embedded Aesthetics."
4 Kwon, "One Place after Another."
5 Berthoin, "The Studio in the Firm"; Berthoin, "Artistic Intervention Residencies."
6 Bastashevski, "The Perfect Con."
7 Rendell, *Art and Architecture*.
8 Wilk, "The Artist-in-Consultance."
9 Rycroft, "The Artist Placement Group"; Dixon, "The Blade and the Claw."
10 See for example, Foster and Lorimer, "Some Reflections"; Mulholland et al., "Resounding."
11 See for example, https://www.ncl.ac.uk/cre/residence/ (accessed 16/1/2020); Gkartzios and Crawshaw, "Researching Rural Housing"; the UK-based Leverhulme artist residencies listed in Table 5.1, the School of Geography, University of Melbourne, Artists in Residence Program, https://geography.unimelb.edu.au/engage/AiRG-program; https://biosphere2.org/institute/artist-in-residence-program (accessed 16/1/2020); http://simeon-nelson.com/index.php/cryptosphere/ (accessed 16/1/2020).
12 Straughan, "A Touching Experiment"; Straughan and Dixon, "Rhythm and Mobility."
13 Grzelec and Prata, "Artists in Organisations"; Stephens, "Artists in Residence"; Higins, "Inspiration and Exchange"; Hercombe, "What the Hell?"; Harris, *Art and Innovation*.
14 Walker, "Artist Placement Group (APG)"; Gertner, *The Idea Factory*.
15 Bishop, *Artificial Hells*, situates APG within the context of socially engaged art.

16 https://www.leverhulme.ac.uk/former-schemes (accessed 16/1/2020).

17 E.g. Lee et al., "Art, Science and Organisational Interactions."

18 https://www.kcl.ac.uk/cultural/Artists-in-Residence; https://geography.unimelb. edu.au/engage/AiRG-program see also https://www.birmingham.ac.uk/culture/ artists-in-res.aspx (accessed 16/1/2020).

19 https://arts.mit.edu/cast/projects/visiting-artists/ (accessed 16/1/2020).

20 https://www.nsf.gov/funding/pgm_summ.jsp?pims_id=503518&org=OPP (accessed 16/1/2020).

21 https://wellcome.ac.uk/sites/default/files/wtx057228_0.pdf (accessed 16/1/2020); see also Glinkowski and Bamford, *Insight and Exchange*.

22 http://artistsinlabs.ch (accessed 16/1/2020).

23 https://www.leverhulme.ac.uk/former-schemes (accessed 16/1/2020).

24 https://www.leverhulme.ac.uk/about-us/history-trust (accessed 16/1/2020).

25 This data is generated from the Annual Review of the Leverhulme Trust, downloaded from https://www.leverhulme.ac.uk/annual-review (accessed 16/1/2020).

26 Ibid.

27 Ibid.

28 This data is generated from the 2001 Annual Review of the Leverhulme Trust, downloaded from https://www.leverhulme.ac.uk/annual-review (accessed 16/1/2020).

29 This data is generated from the Annual Review of the Leverhulme Trust, downloaded from https://www.leverhulme.ac.uk/annual-review, data from 2001 report (accessed 16/1/2020).

30 This data is generated from the Annual Review of the Leverhulme Trust, downloaded from https://www.leverhulme.ac.uk/annual-review, data from 2005 report (accessed 16/1/2020).

31 https://www.a-n.co.uk/news/leverhulme-trust-residencies-plea-resuscitation-important-interdisciplinary-scheme/ (accessed 16/1/2020).

32 https://ecoartscotland.net/2017/03/25/leverhulme-trust-terminates-artist-in-residence-grants/ (accessed 16/1/2020).

33 https://www.a-n.co.uk/news/leverhulme-trust-residencies-plea-resuscitation-important-interdisciplinary-scheme/ (accessed 16/1/2020).

34 Blog. Interview data.

35 http://www.sustainablepractice.org/2017/03/28/leverhulme-trust-terminates-artist-in-residence-grants/ (accessed 16/1/2020).

36 https://ecoartscotland.net/2017/03/25/leverhulme-trust-terminates-artist-in-residence-grants/ (accessed 16/1/2020); https://www.a-n.co.uk/news/leverhulme-trust-residencies-plea-resuscitation-important-interdisciplinary-scheme/ (accessed 16/1/2020).

37 https://www.a-n.co.uk/news/leverhulme-trust-residencies-plea-resuscitation-important-interdisciplinary-scheme/ (accessed 16/1/2020).

38 This data is generated from the Annual Review of the Leverhulme Trust, downloaded from https://www.leverhulme.ac.uk/annual-review (accessed 16/1/2020).

39 See, for example, http://www.meansealevel.net/projects/blue-antelope/_(accessed 16/1/2020); https://www.paulaturner.org/leverhulme/___(accessed 16/1/2020); https://www.ncl.ac.uk/gps/geography/research/leverhulmeresidency/___(accessed 16/1/2020); http://www.matthewflintham.net/blog/2015/1/3/the-martial-heavens (accessed 16/1/2020).

40 Foster and Lorimer, "Some Reflections"; Flintham, "Visualising the Invisibly"; Mulholland et al., "Resounding."

41 For example, Rachel Pain has continued to work with Ribbon Road her collaborators; Flora Parrott went onto do a funded PhD in Royal Holloway Geography and participated in the writing of a successful European Research Council grant application that includes a five-year post-doc for her. Other guy from Newcastle who went onto to a PhD.

42 Evans, *The Myth Gap.*

43 https://geography.unimelb.edu.au/engage/AiRG-program (accessed 16/1/2020).

44 Ibid.

45 Ibid.

46 https://www.kcl.ac.uk/cultural/Artists-in-Residence (accessed 16/1/2020).

47 Ibid.

48 https://www.kcl.ac.uk/Cultural/Artists-in-Residence/About-Kings-artists.aspx (accessed 16/1/2020).

49 https://www.kcl.ac.uk/cultural/Artists-in-Residence_(accessed 16/1/2020).

50 Ibid.

51 Ibid.

52 Tolia-Kelly, "The Geographies of Cultural Geography," p. 135.

53 Ibid.

54 Bastashevski, "The Perfect Con."

55 McLean, "Hos in the Garden."

56 Ibid.

57 Ibid.

58 https://www.a-n.co.uk/news/leverhulme-trust-residencies-plea-resuscitation-important-interdisciplinary-scheme/ (accessed 16/1/2020).

59 Ibid.

60 http://www.matthewflintham.net; http://www.matthewflintham.net/blog/2015/1/3/the-martial-heavens (accessed 16/1/2020).

61 Data from interviews.

62 The Future of Landscape and the Moving Image, https://thefutureoflandscape.wordpress.com (accessed 16/1/2020).

63 https://www.ncl.ac.uk/who-we-are/social-justice/stories/housing-injustice/ (accessed 16/1/2020).

64 Interview.

65 Ibid.

66 Ibid.

67 Ibid.

68 http://www.ribbonroadmusic.com/news/index.php (accessed 16/1/2020).

69 Interview.

70 Bishop, *Artificial Hells*; Rycroft, "The Artist Placement Group."

71 Wilk, "The Artist-in-Consultance."

72 Walker and Latham, "Incidental Person".The Latham archive can be accessed via http://www.ligatus.org.uk/aae/ (accessed 16/1/2020). For an account of creating it see Velios, 'Creative Archiving." The Tate acquired the APG papers in 2004, two years before Latham's death On the APG history and outputs see Hudek and Sainsbury, *The Individual and the Organisation* In these notes, papers from the Latham archive are described with the link to the online record. The APG papers are identified by the prefix 'TGA' followed by the box and file number.

73 Note above.

74 Bishop, *Artificial Hells*, p. 176.

75 Ibid.

76 Hudek, 'The Incidental Person'; TGA 20042/2/1/4/84: Draft of Some Distinctions and Glossary 1978 for The 'Incidental' Person subheading. It is in this 1978 note that the term Incidental Person is first used. The concept itself however is one that can be traced back to Latham's earliest time-based work.

77 Slater, "The Art of Governance," p. 3.

78 Ibid., p. 23.

79 https://www.paulaturner.org/blog/brief-encounters-of-the-chair-kind-aka-why-is-that-there (accessed 16/1/2020).

80 https://www.paulaturner.org (accessed 16/1/2020).

81 Ibid.
82 Ibid.
83 Ibid.
84 https://www.grandgesturesdance.uk/about-grand-gestures (accessed 16/1/2020).
85 Foster and Lorimer, "Some Refelctions."
86 Ibid., p. 430.
87 https://www.paulaturner.org (accessed 16/1/2020).
88 Ibid.
89 https://www.paulaturner.org/blog/prime-position-a-water-bearer-and-a-fabric-folder (accessed 16/1/2020).
90 https://www.paulaturner.org (accessed 16/1/2020).
91 Ibid.
92 Ibid.
93 Ibid.
94 See for example, Hawkins et al., "The Art of Socioecological Transformation."
95 From interview transcript; https://www.geog.ucl.ac.uk/news-events/news/news-archive/2016/august-2016/creative-writing-in-geography (accessed 16/1/2020); http://encountersroom108.blogspot.com/search/label/About (accessed 16/1/2020).
96 From interview transcript.
97 Scalway, "Patois and Pattern"; Driver et al., *Landings*.
98 Driver et al., *Landings*.
99 Parrott and Hawkins, "Six Voids."
100 Henderson, "Academic Conferences."
101 Ibid.
102 Harvie, *Fair Play*; Hawkins et al., "The Art of Socioecological Transformation."

6

THESIS

Prologue: Some recent practice-based PhDs in Geography undertaken in the Department of Geography, Royal Holloway University of London.

في ســوريا، تخفي الكثير من الحدائق جثث
القتلــى من الناشــطين، وتحمــي من بقي
حيا منهم من الضربات العنيفة للنــظام.
في تـــلك **الحــدائق تـــحكي**
GARDENS SPEAK المــدافن
المــنزلية،
Tania El Khoury تانيا الخوري ثمّــة تعاون
مســـتمر بين الأحياء والأموات. الأموات
يحمــون الأحيــاء من خلال عــدم تعريضهم
للمزيــد مــن الخطــر علــى يــد الدولــة،
والأحيــاء يرعــون أمواتهــم ويحفظــون
هوياتهــم وقصصهــم فــي باطن الأرض،
رافضين أن يتحوّل موتهــم جزءاً من أدوات
النظــام فــي التلاعــب بالتاريــخ. الحدائق
تحكــي هــو عــرض فنــي تفاعلــي يجــول
العالــم، ليــروي التاريــخ الشـــفوي
لعشرة أشــخاص دفنوا في حدائق بسورية.
رُكبت هــذه الحكايــات بعناية مــع أصدقاء
القتلى وأفراد أســرهم، لنــروي قصصهم
كما كانــوا ليرووها بأنفســهم. يحتوي هذا
الكتــاب علــى الروايــات العشــر باللغتين
الإنكليزيــة والعربية المحكيــة. مع مقدّمة
للفنانة ورســوم تصوّر تجربة العرض الحيّ.

FIGURE 6.1   Tania El Khoury, *Gardens Speak*. (Image, author's own)

Tania El Khoury, *The Audience Dug the Graves: Interacting with Oral History and Mourning in Live Art* (graduated 2018, co-supervised with Department of Drama, Theatre, and Dance.

This thesis composed four key parts: a performance of the live artwork *Garden's Speak* attended by the examiners during one of its fifteen performances at Battersea Arts Centre (London), an artist's book, documentation of other pieces Tania had created, and a written document of around fifty thousand words.[1] The focus of the text was an exploration of the political potential of interactive live art focussing on the use of oral history and the politics of audience interactivity. It's core case study was *Garden's Speak*, and its evolution from the oral histories given by the families of ten people who died during the Syrian revolution and were buried in their family's gardens, and how its audiences interacted with this powerful work.

Lucy Mercer, Speculative Emblematics: An Environmental Iconology (graduated July 2020, in collaboration with the Department of English and Creative Writing).

The thesis was constituted by a poetry collection *Emblem* and an eighty thousand word written document that together 'comprises a critical and creative investigation of emblems from an ecophilosophical and poetic perspective.' The written document focussed on Andrea Alciato's *Emblematum Liber* (1531), exploring how this 'text-image' artefact enabled the probing of obscurity as a useful concept in the context of the interpretive crises instigated by the Anthropocene. The conclusion to the written text was the only point at which it refelcted on the poety collection, drawing out how the poetry collection explored connections of poetic process, emblematics and obscurity, through early years motherhood.[2]

Nelly Ben Hayoun, 'Homo Faber and Animal Laborans met in Mission Control to Dream of Space: The Design of Experiences at NASA' (graduated Christmas 2017).[3]

The thesis comprised of a forty thousand word written document and two feature-length films, *International Space Orchestra* (2012–2013) and *Disaster Playground* (2013–2015), both of which were presented to the examiners on DVD, the examiners attended screenings of the films at the British Film Institute in London. The written document, which included an appendix detailing Nelly's other work, was focussed on exploring the two films, both of which document Nelly's research and social actions in NASA (USA). It argued that these works were examples of the design of experiences, and that this was a practice and method that combined critical design, critical thinking, and performance to produce research and social actions in situations, thus proposing social and political critiques.

Clare Booker, 'Imagined Airport: a practice-led project exploring the fragmentation and multiplicity of airport sites and spatial experience through methods of collage' (submitted March 2020).

The thesis comprises a body of multi-media practice, including paintings as well as films, and experimental text-based work and a forty thousand word document. The practice is accessed through a specially designed website and an artists' book.[4] While the written document focusses on analysing how three of

FIGURE 6.2   Clare Booker, *Imagined Airport*. (Image, Clare Booker)

the projects presented, *Street View Diaries* (a series of short films), *Apron Space* (short films, drawings, and paintings) and, *Model Fiction* (a virtual airport model that presents a 3D collage of virtual and real space) enables perspectives on themes of spatial perspective, movement and people. In doing so, it argues that this practice-based approach can tackle some of the challenges the airport has posed scholars working on airport space.

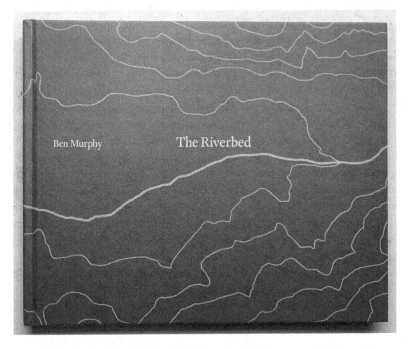

FIGURE 6.3   Ben Murphy, *The Riverbed*, photo book. (Image, author's own)

Ben Murphy, 'A photographic exploration of migrant multi-national countercultural identity through place and space (submitted Autumn 2019).[5]

Focussed on a remote mountainous region of south-east Spain, the thesis comprised several hundred photographs, variously presented in an exhibition at the Architectural Association in Bedford Square in London, an artists' book, and a compilation of the photographs submitted with the sixty thousand word written document. This document used photographic theory to explore how photographic practice – both its making and the resulting images – evolved cultural geographical accounts of marginal communities and complex questions of identity and belonging. Three chapters focussed on how the images enabled explorations of place, home and identity and contested objects, at the site. In doing so it also sought to advance cultural geographies of photography as practice-based research.

Introducing the practice-based PhD

The doctorate has emerged as a key site for the evolution of relations between geographical research and art. As with geographical research more widely, creative practice is made present in PhD research in a rich range of ways. This includes those scholars for whom photography, film, participatory art, or creative writing offers a research method,[6] providing an important source of data or mode of analysis, to those scholars, like those whose PhD descriptions opened this chapter, who undertake what are variously called practice-based or practice-led research PhDs. What marks these forms of PhD out, whether in geography or across the visual and performing arts, creative writing, design and so on, is that creative practice sits alongside a written document that together constitutes the thesis, as such the practice itself is examined as part of the research output alongside writing. It is these forms of PhD that are the focus of this chapter.

In 2001 Fiona Candlin observed that the emergence of practice-based doctorates in the visual and performing arts could be understood as a complex interplay of socio-historical and educational forces.[7] Perhaps most crucially from the 1960s onwards, art schools were increasingly incorporated into the university system. A decade after Candlin's observation, in 2011 it was noted that practice-based PhDs in the arts were still an 'evolving research culture,' with many questions still be answered. At around the same time, leading art theorist and emerging commentator on studio-art degrees, James Elkins observed an exponential growth globally in creative practice-based/-led doctorates, reflecting that while their continued existence was no longer in question, their rigorous theorising was badly needed.[8] These questions continue a decade later still, circling around to issues of judgement and ongoing debates about legitimacy – whether this be parts of the art world that decry the PhD as sell-out institutionalisation, or an academy that requires practice-based scholars conform to publication and funding targets and subscribe to terminologies such as rigour. For others, there are ever-present concerns around the double-standards of a research culture founded on textual imperialism, in other words a culture that is seen to undermine creative practice's claims to knowledge production by almost always requiring the production of an accompanying text to credential the practice as research.[9] This chapter draws on the expansive discussion of these forms of PhD which exists across visual and performing arts, creative writing, and design whilst focussing on how practice-based forms of doctoral study have been playing out within geography. As such it will complement the vast majority of published accounts of practice-based PhDs (the term this chapter will adopt) which focusses on arts-based disciplines, and it will extend these discussions. It will respond to desires to move away from the naming debates, and to provide studies of the experiences of the doing of these PhDs, rather than focusing on accounts of the work they produce.[10] This chapter contributes to this wider body of work by foregrounding the spaces, practices, and experiences of practice-based PhD students within geography as well as the experiences of those who supervise and examine them. In doing so, it both builds on and seeks to extend literatures on practice-based PhDs by first offering perspectives from a non-practice-based discipline and, second, by way of its geographical perspective on these forms of doctoral work.

There is significant global and disciplinary disagreement and policing with respect to the terminology for these doctorates – practice-based, practice-led, practice-as-research, research through practice, creative practice as research, artistic research, performance as research.[11] More than simply globally unfolding academic turf wars, these debates often get to the heart of wider concerns about the nature of these forms of work, the relationship between practice and other forms of research, and the legitimacy of the claims that practitioners make to conducting research through their practice.[12] Here I choose to use practice-based, because while acknowledging its contestation, it seems to be the term most commonly chosen by geography PhD students working in these ways. Whatever the terminology used across disciplines, what most clearly unites these disparate ways of working, and demarcates them from other doctoral forms, is the centrality of creative practice to the entire research process, and the belief that the assessment of the creative practice makes up a significant portion of the work. In some cases, creative practice sits alongside a series of publications or a written document of up to 100,000 words; for others a written document of between 30,000 and 40,000 words is more normal; in a few institutions the required written component is zero.[13]

The PhD is, as this chapter will explore, not just a growing site of the coming together of creative practice and geographical research, but also a very illuminating one. To situate art as research within the geography PhD is to insert these practices into the heart of our academic work, 'at the point at which we produce and reproduce the academics of the future, new kinds of researcher with new expectations about what research is, how it will be assessed and what counts.'[14] Indeed, a doctorate is a right of academic passage, and represents the acknowledgment of an original research contribution and a professional level of research competence – a kind of licence to practice research as one of my colleagues describes it. As many commentators on artistic PhDs have noted, it is perhaps because of the doctorate's centrality to the academic system of recognition and judgement, that the introduction and growth of these forms of PhD have been such a point of controversy and anxiety for institutions, disciplines, and individuals – supervisor and student alike.[15]

Interestingly, an insistence on practice is in the DNA of doctoral qualifications. One history tracks the first doctorates to ninth-century Islamic Madrasas, where a successful student would practice religious law under his master until he was deemed experienced enough to be granted a 'ijazat at-tadris,' a licence to practice religious law.[16] Another history situates the emergence of the doctorate in the European Middle Ages, with the first awarded around 1150 in Paris, again, the focus was on practices of theology, law, literary arts, and medicine, with successful candidates awarded a licence to practice and teach.[17] It was not only the nineteenth-century transformation of the University system by Wihelm von Humboldt that original research was situated at the core of the PhD.[18] These doctorates were awarded first to students of philosophy, hence the title Doctor of Philosophy – *Philosophia Doctor*. Many histories of the modern doctorate note a relatively fixed form, either the American model with course-work, exams and a dissertation, or the European Humboltian model, which involved a thesis and a viva.[19] Commentators on the contemporary doctorate place a marker post-1960 when the downturn in the growth of higher education, the decline in state-funded education, and the rise of HE marketisation forced the evolution of the PhD. This included the emergence of PhDs by publication, the professional business doctorate – the DBA, as well as the practice-based doctorate for creative practitioners.[20] If the histories tend to suggest a replacement of practice by original research, then, in its current iteration, the creative practice-based PhD insists that practice *is* original research.

The focus of this chapter is the still evolving field of these practice-based PhDs within the discipline of geography. The more general rise of practice-based PhDs is dated to around the end of the twentieth century and early years of the new millennium. In 2007, John Hockey observed that despite their critics over forty UK Higher Education Institutions would grant these PhDs.[21] The origins of these doctoral forms in Geography is perhaps less clear.[22] Not least because some candidates, such as Perdita Phillips, gained their PhDs in Fine Art but in what are avowedly geographical research topics. In Perdita's case her practice-based doctorate (awarded in 2006, from Edith Cowan University in Australia, in Visual Arts from the School of Communications and Contemporary Arts) focussed on the contemporary art project *fieldwork/fieldwalking* that explored practices of walking and science in the field.[23] I am especially keen here however, to look at practice-based PhDs taken largely within (although alongside other departments and supervisors) and awarded by, Geography Departments. At Royal Holloway University of London in 2013 we did a global search for Geography Department guidelines on awarding these forms of doctorate. We found no disciplinarily tailored guidelines, but did see evidence of a few cases where practitioners working in geography had used regulations developed by practice-based disciplines elsewhere in their institution. It was clear that a series of geography PhDs had incorporated the production of films, bodies of photography or creative writing as part of their data collection. We struggled however, to find any that would be considered to be 'practice-based' as visual and performing arts disciplines understand it. In other words, where practice comprised a significant part of the thesis accompanied by a written document. A global survey of geography departments in 2019 estimated that about fifty or so of these practice-based PhDs had either been passed or were in progress. This evolving research culture is explored here through research with completed and ongoing students, as well as with supervisors and examiners, and my own autoethnography as a supervisor of ten of these PhDs and an examiner of the same number, informs the following discussion. The chapter moves through a series of spaces associated with the PhD: the supervision, the thesis itself, and the viva. As well as offering an account of these important forms of interaction between creative practice and geographical research, exploring these ways of working directs us to questions of judgement and disciplinary competency, and to issues around writing, theory and practice, and the all-important question of text and practice relations.

The supervision

In many writings on the supervision of practice-based PhDs, the supervision emerges as a site of anxiety for students and supervisors alike.[24] Probing this anxiety further in terms of my own experiences and those of other geography supervisors, it seems that one of its sources is the fundamental questions about the PhD, its nature, use, format, and processes, including that of supervision, that emerge. Within geography, these PhDs also seem to become powerful lenses through which to reflect on the adequacy of our disciplinary research cultures and practices to support practice-based working. Particular points of tension and flex emerge, and many of us become painfully aware of the need and our ability (or sometimes lack thereof) to negotiate on behalf of our doctoral candidates. The experience of supervisor and student often seemed to be one of confronting over and over again challenges both of the everyday and mundane practices and bureaucracies of the institutionalised PhD process (where to hold supervisions, what work is set, what is expected for review processes), as well as far less mundane elements such as thesis format and the nature of the viva.

Supervision is a highly varied process, whether of practice-based PhDs or not, so it is no surprise that supervisory practices varied considerably across geography. The basic constitution of the team often varied, not only with the nature of supervision across countries and across institutions, but also dependent on the nature of the thesis being undertaken and the experience of the practitioner. Our normal practice at RHUL, and this seemed to accord with other geography departments around the world, was to ensure a supervisory team that had some practice-based experience present as co-supervisor, or advisor, as part of the committee, the exact form and title given depending on the terminology used at the institution. Oftentimes, myself and other supervisors found ourselves building exciting working relationships with scholars in very different parts of the university as a result of co-supervising these PhDs. Within my own institution I have worked with creative writers and drama scholars, leading to further PhD supervision. Others described how this process built new research relations, kick-started artists-in-residence programmes, and even brought geographers closer together – across our sometimes internal disciplinary divisions. Amongst my own students, I have had several who had no supervisor from a practice-based field. These students tended to be professional practitioners with many decades of practice under their belt, who had come to geography seeking a particular conceptual framing for their work, rather than to focus on evolving their arts practice.

There has been a growth of literature recently on the skill-set of the doctoral supervisor.[25] Alongside intellectual knowledge, the focus has been on social skills and qualities of empathy and compassion. Talking with practice-based students within geography it became clear that to this list should be added, an experience and awareness of practice. Many indicated the value and importance of geography as a location for their work and hence the support of a geographer was a crucial part of their PhD experience. As one noted, this had been important for 'giving me geographic chops, I now call myself a geographer,' and indeed this individual now holds a permanent post in a geography department. For other students however, a series of struggles emerged as a result of a lack of regular contact with practitioners – often where there was no practitioner on their supervisory team. One noted that, despite well-meaning supervisors, geography as a discipline needed more 'professionals who can adequately support a creative practitioner.' Often cited by this student and others, was the need for supervisors to understand working practices, including appreciating the challenges of bringing together practice and theory, the challenges of writing as a practitioner, as well as ideally more help in the negotiation of institutional regulations and expectations that did not always make space for practice-based working. Another aspect that repeatedly came up in discussion was the challenges students faced around being professionalised or encultured into a research culture. My own experience as a supervisor made it very clear how much I had taken for granted the networks and contacts I could offer those PhDs I worked with. Whether that be links with more senior scholars, advice on conferences, publication, careers or access to workshop space, cheap resources, or gallery space for exhibitions. Interestingly, some students found that their practice-based nature of their work evolved during their PhD candidature, and at one institution it was possible to find an additional supervisor at the affiliated arts school to join the team, and joint candidature of both departments was agreed. If some students expressed concerns around the lack of practitioners to support them within geography, another reflected that the practice-based background of a supervisor was seen as less important than fostering a 'dynamic relationship between supervisor and student where skills are developed in relation.'[26]

As well as specific supervisory skill-sets another point of discussion for practice-based geography PhD candidates was the research culture that surrounded them. In several cases, the location within a geography department was a choice driven by an active desire to not be in a visual arts or poetry department. One PhD candidate, an established artist already teaching in arts universities reflected on her choice to come and work within geography,

> I realise that I have a very different approach to other students and academics working in practice based/ led – but there is somehow expected to be a common language with for example a performance artist, sci-arts or creative writer- in fact I often find I have more in common with PhD students from outside of creative methods groupings.

For others, often those earlier in their careers, a commonly articulated concern was around the need to feel immersed in a practice-based research culture. As one geographically located student noted, on having negotiated access to arts department activities, 'for me, it was less about generic training, and more about informal training and discussion- being immersed in a culture of practice-based research attending seminars and exhibitions.' Often as a prospective supervisor I found myself apologising to students about the lack of access to materials, space to work, and other facilities (dark rooms, rehearsal space). Yet what existing students observed was less that a lack of studio-based work culture led to a lack of material resources, and more, that it led to the absence of shared discussion of the work in progress. Often called 'crits,' these apocryphal art school rites-of-passage are beloved by some, and hated by others, literature goes back and forth regarding the focus of attention – less a point of judgement than an 'attempt to get into the inner logic of the work'. Should it be the form and materiality of the work being presented, its influences, or what the artist says about the piece during their introduction of the work to the group? Oftentimes, it seems that these all play a role in the discussion. Loved or loathed, it is clear that the 'crit' is a key part of how artistic practice evolves for many of those based within arts institutions and is a vital method of teaching practice.

The process of supervising practice-based PhDs within geography, and talking to other supervisors and students, attuned me to the myriad ways that geographical, and oftentimes wider institutional, research cultures don't always suit, let alone support, practice-based candidates. Indeed, to properly support these forms of doctorate within geography required me and others to be willing to revisit and be open to challenging the standard biographical process of the doctorate. This included the nature of the initial contact, the myriad intricacies of applying for funding – from form filling, proposal design and CV maximisation to the selection of target funding bodies – not to mention the actual research itself and its intersection with the regulations of institutions. Practices of supervision, modes of engagement with ongoing research, formal institutional milestones, exceptions and regulations, as well as the form of the thesis itself and the viva, were often put into question. This was invigorating but also at times wearing; it sometimes felt like having to 'go into battle,' over and over again, to ensure that what practice-based students in geography were being asked to do was compatible with working methods and that disciplinary 'norms' where not putative, underplaying, or even just overlooking strengths. At the point of application, for example, a practice-based scholar's CVs might look very different, and working out how to present these in a way that ensures recognition for exhibition and practice by panels not used to such features is important. To be in a position to be able to hold a discussion with assessors is rare. Further, the research proposal format will likely be alien to a practice-based researcher, with few doing the undergraduate or masters dissertations that introduce most geography students to the form. Another issue concerns key institutional milestones, practice-based candidates likely produce less written work, as they will also be developing their practice, yet guidelines for submission for annual reviews and upgrades can cause concern. Further, something as apparently simple as the supervision location also needed more careful consideration –effecting the nature of the discussion that would be held. My normal practice of meeting in offices, coffee shops, or the occasional seminar room was not really adequate. Further my *modus operandi* of setting a writing task, receiving the work, and then meeting to talk about it also needed some careful rethinking. At some points in the research process it worked fine, but if I was to respect the role of creative practice in the doctorate then more space, literally and metaphorically, need to be made in for practice in supervision. As one PhD student (not one of mine) located within an art school observed, 'I think it is fair to say that probably 90% of the formal discussions I had were about the status and value of the written component. This was necessary but unfortunate.'[27] Supervising geography practice-based PhDs meant ironically, not overlooking the importance of geography. If work was to be looked at, and the institution lacked studio space, students could not be expected to transport all the work, sometimes delicate sculptural forms or large-format photographs to the supervision, nor would pictures on a laptop always be adequate. Nor could I always set work that would result in writing. As such, studios become supervision sites, as did screening rooms, performance spaces, and the landscapes that had inspired the works. I got used to responding to material sent ahead of supervisions that is not just text I can easily call up and annotate on screen. Instead, feedback required time spent engaging with artworks, drawing on all my training as a geographer and an art theorist to offer reflections on videos, sculptural work, paintings, and so on. How I responded also became a point of discussion. As several students politely noted in various ways, while my questions were interesting they were often different in focus than those asked in a studio context. Informed by geography and art theory, I might query links between the form of the work and geographical literature on say experiences of underground space, I might probe how the mediated experience of a piece of work for its web-based audience related to the conceptual discussions about say non-linear experiences of space. These were however, firmly conceptually focussed questions. Rarely did I ask questions about the materiality of the work in and of itself, why was something only so big, why not huge? Why was it fabricated in this way rather than that? I could not give advice about workshop practices, or identify where technique could be refined, or suggest other making methods. In some cases, these were eased by my role as part of a supervisory team in which there was a clear division of labour. In the case of Lucy Mercer (a description of whose thesis was included in the prologue) I co-supervised with the poet Jo Shapcott, a professor of creative writing at my institution. Lucy sent Jo her poetry, and would send me what she called the 'critical' text on Alicato's *Emblem Book*. 'Creative' and 'critical' in Lucy's case were thoroughly enmeshed (her poetry being highly critical, and her environmental humanities analysis being highly creative) but as terms they served as a useful shorthand for the division of supervisory labour. We would come together to view the thesis as a whole at key milestones such as annual reviews. This division of labour would, of course, not work well for every supervisory team, every topic, and all iterations of the relationship between practice and text. It is to this relationship between practice and text that discussion now turns.

The thesis as medium: assembling portfolio and page

Where a student working in the field of Geography has undertaken research in which practice forms a core methodology and mode of research, the PhD submission may include a body of creative work devised specially for the degree together with a body of critical/theoretical writing. Both will show coherence, originality, and intellectual rigour, illustrating knowledge and understanding of relevant practice and critical debates in the field. The relationship between the creative and critical/theoretical components will vary depending on the specific project. Taken together, they will demonstrate the contribution to knowledge made by the thesis as a whole. The exact balance between the creative and critical elements of the thesis will be decided between the student and the supervisory team; however, the following is a suggested breakdown:

- *A substantial body of creative work;*
- *Normally 30,000–60,000 words of critical writing.*

> *If an exhibition, live performance, or other non-textual creative work is to be submitted as part of the PhD, it will be appropriately documented and a retainable record of these creative practices will be submitted in a form which has been approved by the supervisory team and the Director of Graduate Studies in Geography.*

An extract from Royal Holloway University of London
PhD regulations for practice-based work within geography

What a practice-based PhD actually consists of is one of the enduring questions of the field. From the art-world perspective, core to this question is the

controversial concern with how creative practice is made present in the submitted and examined doctoral thesis.[28] Principally, the question is one of the
relationship between the written document and the body of creative work that is
presented for examination. The creative practice can of course take a wide range
of forms, and needs to be presented in an appropriate way for the examiners,
but also in a way that stands the test of time – the thesis needs to be complete
for audiences that might later consult it. As even just the PhDs included in this
chapter's preface makes clear, the kinds of practice that PhD students might create is myriad, from performance works, to poems, photographs, paintings and
films, each of which poses different challenges in terms of how to mediate the
experience of examiners, but also of later audiences. These challenges might
include where to stage and how to document a live performance or exhibition,
often with limited funds; how to enable later access to photographs and paintings best seen at large scale; or how to curate the experience of a series of films
for examiners and later audiences. A central relationship to consider is that between the creative work and the written component. The latter, in some cases
by-publication, is also hugely varied in both length and content. Some produce
around 30,000 words of text, often exegesis on the creative practice that is its
subject, directly addressing how it tackles the research questions to hand. Other
longer written documents situate the practice within wider debates offering a
broader discussion and contribution to the research topic, similar to that which
one might expect from a non-practice-based thesis. At either extreme are those
who choose to write a thesis of 80–100,000 words long, which addresses an allied
topic to the practice but is not focussed on the practice per say, and those who
submit no written document at all. Currently the International Interplanetary
consortium, which awards Art-Science PhDs from The University of Plymouth,
UK, supervised at art institutions around the world is one of the only locations
offering practice-only doctoral candidature.[29] This practice-only policy is often
described and valorised in terms of one of the dominant circulating imaginations of the written component – that it is 'double work.' In other words, if
artwork is taken seriously as knowledge then why is a written accompaniment
also required?

Yet my work with geography doctoral candidates and research on their thesis, suggests that they have been negotiating this relationship between portfolio and page perhaps a little differently. I use the term 'portfolio' here to cover
the practice component of the submission, which includes 'finished' work as
well as documentation of work in progress or past exhibitions and performances.
Page, as it is used here, refers to the written element of the thesis submission.
The imaginations of this written element that circulate through the geography practice-based PhD community both echo, but also deviate from, those
found within the wider practice-based PhD community. As I will discuss, while
the concerns around textual imperialism that shape many wider discussions of
practice-based PhD are very present for geographers too, there are also examples
where the relationship between portfolio and page is being reshaped. Drawing

on ideas from experimental publishing practices as well as performance and live arts understandings of documentation, I reflect on the idea of thesis as medium. Rather than situate the relationship between writing and practice as a negative one, detrimental to practice, I explore how in the case of many practice-based geography PhDs, what seems to emerge is the thesis as an assemblage of creative practice, writing, and even the curation of the viva.

Imagining writing

The writing process is often one of the most tension-filled elements of the doctoral experience.[30] Indeed, in recent years there has been a proliferation within geography of accounts of the writing process. Whether these are the writing struggles of graduate students or senior colleagues alike, the evolution and discussion of various forms of on and off-line writing support groups, as well as a vast expansion in the understanding of what 'counts' as geographical writing practices.[31] Against this context, I assembled a writing group for my own PhD cohort, of which around half – so anywhere between three and six depending on the year – were practice-based scholars. The group aimed to make writing a positive site of exchange between peers, practice, and non-practice-based candidates alike. Sometimes we workshopped chunks of individual's text, other times we debated so-called 'academic style,' what it was and how much could it be deviated from. Other times, we shared varied texts we had found inspiring – both 'academic' and 'non-academic.' It was clear that across the group writing was a deep-seated source of anxiety. Listening to the group share understandings and approaches, the variety of imaginations of writing and where it sat within the research process became clear.

Perhaps least expected, were the views of a few practice-based candidates who displayed a deep-seated rejection of writing as part of their research process. Writing emerged as necessary evil, at worst a waste of time being an outright distraction from the business of making artwork, at best a hoop to be jumped through as quickly and painlessly as possible. This was in contrast to those within the group for whom writing was a daily practice, fundamental to how their research proceeded, or those for whom, over the course of the PhD, it had become a craft. Indeed, some practice-based students, within this group and beyond, clearly articulated how during the course of the PhD writing emerged as a creative practice in its own right. There was pleasure expressed in developing academic writing skills, for some an unexpected, but happy, outcome. For one candidate the practice and the vocabulary for describing it were honed alongside each other, useful not only for the thesis but far more widely. For others, evolving writing skills were a source of excitement and satisfaction as new geographical audiences were opened up for their work; 'I have alot more experience writing for academic audiences, and have been able to reach new audiences with for my work.'[32]

Listening to the discussions, and rejections, of writing. It became clear that these were not just extreme examples of the frustration that often surrounds writing, but also a sense that the relationship between creative practice and written document can be highly ambiguous, but also for some frankly non-sensical.[33] Looking at literature on practice-based PhDs, including student surveys with significant sample sizes, there is a common belief that the written component of the thesis, or any written explanation of practice-based research, is effectively doing 'double work.'[34] In other words, practice-based students develop research through their medium and their final pieces of artwork are the site of the communication of these ideas. So why, in addition, do they have to also communicate these ideas in a written document? There are, as others have hinted with respect to wider creative practice-based research, shades of imagination of the Derridean supplement at work in this relationship between creative practice and text. In other words, that linguistic ambiguity that is at the heart of Derrida's idea of the supplement – with its combination of supplémenter, to add on to, and suppléer, to substitute.[35] For Derridean it is often undecidable whether the supplement adds itself and 'is a plenitude enriching another plenitude, the fullest measure of presence' or whether 'the supplement supplements… adds only to replace… represents and makes an image… its place is assigned in the structure by the mark of an emptiness.'[36] Indeed, for Derrida, writing is already a 'supplement *par excellence*, since it proposes itself as the supplement of the supplement, sign of a sign, taking the place of a speech already significant.'[37] To think then of the text as supplement, is to think of it as both having potential as an accretion but also concerningly, the potential to be a substitution for that practice.[38] This reminds us that the relationship between practice and text is hard to resolve, that it needs further exploration beyond situating the text as somehow representational of, standing in for the object, but that could be easily mishandled.

The complexity of the relationship between text and creative practice is not just a quirk of the PhD, but an enduring issue with practice-based research more generally.[39] Indeed, high-profile gate-keepers, including in the UK, the Arts and Humanities Research Council (the primary research funder for practice-based research) and the Research Excellence Framework, both stipulate that practice should be accompanied by text if it is to be considered research. Thompson quotes the UK Arts and Humanities Research Council as saying, 'creative works, no matter how highly esteemed, cannot themselves be regarded as outputs of research. They can only become so in association with explanatory or contextualising text.'[40] Whilst feedback from one of the panels that dealt with practice-based research in the 2014 UK Research Excellence Framework noted that too often the accompanying texts sought to enumerate the 'goodness' of art through citing reviews, gallery profiles, or visitor numbers. However impressive these might be, this did not in effect prove that this was 'good' research, and that the accompanying text was instead required to 'prove' the work was 'research.'[41]

Much has been written around the need, especially when it comes to the form of practice-based PhDs, to at least acknowledge, if not challenge what is called, after Felix Guattari, research's linguistic imperialism.[42] I understand linguistic imperialism as it is used here to mean the power evoked by language – spoken and written words – over and against non-semiotically formed enunciations, such as the knowledge claims made by a piece of sculpture. Imperialism, clearly a loaded term, speaks here of the nexus of power-knowledge and the hierarchal and subservient relations that Guattari determined as being unevenly distributed across modes of expression, from language dominated by semiotics to forms of affective and emotional expression that might work through other means. Guattari found in art, amongst other aesthetic practices, ways of working and communicating that might topple the written form, bringing into question its particular claims around representation.[43]

To reflect on these discussions of the relationship of text and practice from a geographical perspective adds another dimension to them. Deiter Lesage, a long time commentator on these forms of doctoral work, suggests that these textual requirements might indicate 'either a lack of faith in the capacity of the arts to speak in mindful examples and critical ways in the medium of choice, or in the university's own capacity to make sound judgement on meaning, complexity and criticality of the outputs as such?'[44] I think in the case of practice-based PhDs situated in geography it is perhaps the discipline's capacity to make sound judgement on meaning, complexity, and criticality of the outputs, that is in question. Whilst several generations of cultural geographical scholarship demonstrates that the discipline does take artistic practice very seriously as a form of knowledge making – as offering, in Lesage's words, 'mindful examples' and 'speaking in critical ways' (although note the linguistic metaphor).[45] It seems, however, important to acknowledge our collective disciplinary limitations in terms of passing sound judgement on creative outputs. After all, whilst within cultural geography there are those who work as literary, music or art theorists and practitioners these skills are not necessarily wide spread. In other words, relatively few geographical scholars would possess the training to fully appreciate creative practices as knowledge production. Indeed, as an editor who seeks reviewers for critical work, I often find reviewers responding generously, but, as with my own responses, they are very aware of their limitation in the face of some mediums. It is also important to note however, that those doing practice-based PhDs in geography are very aware of the importance of text as a means to reach geographic audiences with their work. Returning to some of the reflections offered by practice-based PhD students on their experiences, one of the common reasons they gave for valuing the honing of their academic writing skills was in fact to reach those geographical audiences. Without wanting to downplay the frustrations many feel at the centrality of the written word, it seems that those working within geography do realise and value the need for text. Indeed, as I want to go onto explore, some practitioners are not only embracing the production of the written document, but also seeking to experiment with it in ways that build other imaginations of the relations between text and practice other than those of textual imperialism and the dangerous supplement.

Thesis as medium?

One of the very productive sets of discussions that emerged from the writing group was a discussion of how writing could be 'different' than the imagination of 'academic' writing which seemed to haunt many within the group. How to sound academic, was it 'academic' enough was a common refrain across the group. We started to gather examples from geography and elsewhere of different forms and styles of 'academic' writing. The result was a rich discussion that meandered through a number of the sessions about the nature of academic writing and how it could be done differently (some of these threads are taken up further in the following chapter). If one of the reoccurring themes of this discussion was what it might mean to write in the first person and to bring emotions and affect to the page, to write to move, another, often followed up in supervisions, was how in practice-based work the relationship between text and creative practice could be evolved differently. What I saw emerging was a mind-set that saw a more productive relationship, one that began to play with thinking from experimental publishing practices and performance and live art, to explore what might be possible within the context of the 'text' element of the PhD. Importantly, it was not just the possibility to play with content that emerged from these discussions, but also a concern with the form of the 'text' element of the PhD. As such, the need to produce certain forms of text – an introduction, a methodology, a literature review, the discussion chapters, and a conclusion – together with the requirement for a 'bound' A4 volume of text, came to function for some as productive, generative constraints. Candidates variously evolved new forms of curated website, to enable their work to be accessed alongside text in innovative ways. Another attended book-making courses to experiment with binding and formats to enable the production of book as part of their thesis. A further student drew on already existing skills of book-making (as the following chapter will explore experimental publication practices are a common medium in twentieth-century art), to think through how the thesis might respond not only to the demands for content, but might also be a form of publishing experiment.

The history of performance and live art offers a rich location for thinking through problematic relations between creative practice and textual documentation. Theories of live and performance art have long been shaped by a debate about the role of documentation of these 'original' sometimes singular time and space specific performances/events. There has long been a divide between presence and absence models of performance. In other words, there are those who believe in the importance of being present at a live experience, and thus situate the experience of secondary audiences, through videos, texts, and images as a lack. This latter is the absence model of performance, a key proponent of which is Peggy Phelan who considered the ontology of performance to be one of disappearance.[46] A presence model of performance by contrast, suggests that far from being encroached upon, continued, or threatened by mediation, live performance is in fact defined as live through the possibilities of its technical and print mediation.[47] In short, the materials used in representing the performance not only guarantee our continued access to the work but are in fact responsible for qualifying it as performance in the first place. Transposing this argument to discussions of the practice-based PhD, suggests that perhaps the written element might be conceived of as mediation, so rather than being separable from the practice as research, is in fact a part of how it is that practice can be research – part of its ontological being.[48] The written element of the thesis, instead of being a dangerous substitute, something that suggests a lack of understanding and appreciation on the part of a research world, could in fact be understood as something that is integral to this being practice-based research, especially in a discipline like geography. As a tool of mediation the written document takes on a different role and can realise its own aesthetic force, directing the reader to experience another body of practice in a certain kind of way, rather than being a secondary representation of that practice.

Given the relatively low numbers of practice-based PhDs in Geography, in comparison to the many thousands elsewhere, it would be too ambitious to propose this as another model, but it does seem that a case could be made for considering the thesis as medium, rather than only conceiving of the written document as dangerous supplement. As such, the elements of the thesis are part of a wider assemblage of practice that situates the artefactual form of the written document as a kind of bookwork, alongside performance events, alongside sculptural artefacts and installation environments. For some practitioners this might make sense, especially those who are interested in and able to access new or existing writing or bookwork skills. But this might not be the case for all practitioners.

Here practice and writing are folded together as part of the research, rather than either being situated downstream of the other. Importantly though, the purpose of these discussions is not to model one ideal way to produce a practice-based PhD in geography, one singular relationship between written document and practice. Indeed, perhaps the only thing that can be observed is that each and every practitioner I worked with spent a long time reflecting on the possibilities of practice and text relationships. Ultimately, it would be inappropriate to promote one form of relationship over another, but it is important that this is a point of discussion and reflection to those supervising and examining in this area, as sometimes it can seem to slip away in the need to get 'the thesis' whatever that might be finished.

The viva: examination experiences

the PhD submission may include a body of creative work devised specially for the degree together with a body of critical/theoretical writing. Both will show coherence, originality and intellectual rigour, illustrating knowledge and understanding of relevant practice and critical debates in the field.

Extract from RHUL PhD award guidance

My experience of examining creative practice-based doctorates has been diverse and has included those done within and beyond geography. It has covered mediums as diverse as textiles, a range of forms of visual art, documentary
film-making, performance and live art, design, dance, and geopoetics. I have
participated in vivas where the thesis included very little text and it was clear that
the practice was central to the research. I have also participated in vivas where
had I not queried it, the practice might have been totally missing from discussion
and reflection, dominated as the interaction was by the written document in front
of us. Oftentimes, as with those vivas I have organised for my own students, I – as
a geographer aware of practice-based research – was examining in combination
with a practitioner who might have an expressed interest in the medium and less
often the research area being explored. As such, as a team we combined competencies in the geographical (or in some cases sociological) research topic at hand,
and the practice. At times though, I found my competencies stretched. When, for
example, I was the geographer-who-does practice, paired with a sociologist or a
geographer with little or no experience of practice, or when I was being asked to
work outside my geographical area of expertise. In other words, when I had been
recruited as a geographer who was friendly to practice, rather than a specialist
in the research topic per say. Such occasions caused, unsurprisingly, significant
reflections on what it meant to be asked to judge practice-based research and
on what it was that I was looking for from the practice in combination with the
written document. At some points, my role as examiner has been to assert the legitimacy of practice-based research in the context of geographical topics, to help
ensure that the work the student has produced is valued for its original contribution through practice. I have encountered written texts that focus on practice, yet
at no point helped me appreciate how I should encounter the practice. Should, I
view all the video links ahead of reading the text? Should I view the video discussed in the chapter only ahead of reading it, after or during? Is it up to me how
I encounter the text, is there any curation by the candidate?

The root much of the anxiety that pervades literatures on the practice-based
PhD appears to be questions of judgement and uncertainty around these PhDs
and their validity.[49] Fiona Candlin observes how 'the specific criteria of competence for the practice-based PhD is [are] not therefore immediately obvious...
how do you produce or examine a PhD when it is unclear what competence
constitutes per se?'[50] While it is clear that art schools and contexts of art critique
have forms and standards for evaluating art, it seems that such forms of evaluation
do not always translate well when we move artistic practices into the academic
contexts.[51] A key site where questions of judgement and evaluation come to a
head in the PhD process is that of the viva. Yet if there is much controversy about
the written components of the PhD thesis, there seems perhaps surprisingly a lot
less around the PhD viva. Many of those students, supervisors, and examiners
I spoke to, and most of those I supervised and examined, had at least two components to their viva. Across the world geographical viva practice varies, from
two examiners and the candidate (common in UK Geography) to committees

of multiple people either in private with the candidate or in front of audiences of colleagues and family. Oftentimes with practice-based work however, there is an added component to the normal discussion: in that there is an exhibition to view, a film screening or live performance to attend. In short then, just as the practice-based PhD demands we re-visit the written component of the thesis and ask questions about its role, so we are also required to reflect on the viva, its form and content, the ways judgements are made, and importantly, how practice is presenced during this process.

One of my core learnings from collaborating on practice-based supervisions with practitioners was to start reflecting early on how the examiners are going to encounter the creative practice. This needs to occur early on in the PhD process, and certainly while the form of the written document is still open. Such a process constitutes the thesis almost as an act of curation, where the encounters with the creative practice, written document, and experience of the viva must all be configured together. There are both practical and conceptual elements to this process, what can be afforded financially in terms of showing the work, what can be practically organised in terms of timing: many good spaces for performances and film screenings book up years in advance, and geography departments don't always have easy access to appropriate spaces. What is also important is to think through the ordering of the encounters, does it matter if the examiners read the text after the film screening, or should they see the live performance before they receive the text?

How will practice, especially it is temporally and spatially specific practice – live art, site-based work – be made present during the viva, will artefacts be brought into the room? Might the viva be held in an exhibition space, including an exhibition tour, will the practitioner be present? In some cases these kinds of queries are not needed or relevant, or the answer is obvious, dictated clearly by the nature of the work or practical and conceptual issues. In other cases these decisions are important points of conceptual reflection that speak powerfully for student, supervisor, and examiner to the ways in which the relation between practice, writing, and research is being performed and understood. To give an example. On encountering the written component of one individual's thesis ahead of the viva, I was puzzled. The document contained some rather ordinary photographs of the sculptural and drawing practice that was the practice component of this practice-based thesis. These images felt more like illustrations than a display of practice: many were in appendices; some lacked title, size or details of medium; some were badly lit. I concluded from reading the document that an exhibition of the work at the viva was being understood as the primary site for encountering the practice. Yet on arriving at the venue of the viva, this turned out to be quite a small exhibition, in a seminar room, with tables as plinths. We held some discussion with the candidate in the exhibition, and then moved to another smaller classroom for a longer discussion with the written document in front of us. It became clear during this discussion that there was a definite rationale for the logics of work's presence in the exhibition and the written document.

The candidate did not accord the practice of display and the idea of a 'finished' work with any status, rather he conceived of his practice as an ongoing material and conceptual churn – so the 'display' of a finished, final piece would have been inappropriate. Once this became clear it was possible to reflect with the candidate on this, and how and why it might be appropriate to make the aesthetic decisions clear in the written document for those who encountered the thesis later on. In an almost diametrically opposed example, another PhD candidate expressed his desire to curate a site-specific viva for his examiners. As such, he wanted the examiners to have the experience of coming to the site, and moving through it, before encountering the work in situ, as it was designed to be consumed. This, the candidate believed was integral to how the examiners should experience the work. A further question then became, how would this experience be recorded and/ or replicated in documentary form for audiences at a later date. Such a curated experience would not be appropriate for all PhD candidates; indeed it would only make sense if the form of the viva resonated in some way with the research itself.

Aside from the importance of thinking through the constitution of the examination process, the viva is also interesting for the point of judgement it offers on the PhD and the questions this inevitably raises for those of us involved in these processes. Reflecting on the use of peer-review panels (detailed in Chapter 7), Jennifer Wilson observes the mismatch between what she terms the 'black box' of peer review criteria around questions such as rigour, appropriate critical context, and the 'white cube' of artistic making.[52] As the section from the criteria for practice-based PhDs in RHUL Geography above makes clear we have a similar set of criteria. We expect both the body of creative practice and the written document to demonstrate, 'coherence, originality and intellectual rigour, illustrating knowledge and understanding of relevant practice and critical debates in the field.' Attention to evaluation within practice-based circles tends to be directed towards the danger of accretive expectations of the PhD, querying what exactly we expect the research to demonstrate. Are we asking that practice-based PhDs offer a transformative practice in that 'art in short changes art?'[53] Artistic research is therefore an advancement of artistic practices and needs to foreground artistic novelty.[54] In the case of geographical practice-based PhDs are we looking for a contribution to geography, are we looking for practice-based research that transforms geography? The risk seems to be that either we end up asking too much, to transform geography and transform art, or that we risk the production of mediocre artistic research in artistic terms, but good geography, and does this matter? Yet it seems that for many of the practice-based PhD students I spoke to, to develop good art and good geography could be the same goal. In other words, the practice that contributed to geographical knowledge is also a practice that advances art. Indeed, it seems for some practitioners it was often the very engagement with geography that was understood to be what would evolve art.

For many of those I interviewed, entry into a geography PhD was a deliberate choice to offer a ground for furthering artistic potential based on the chance it offered to develop work through the reading, writing, and thinking associated with 'research' and with theory. This is in tune with wider studies of arts PhDs which have found these forms of doctorate to enable writing and reading and other practices that are seen as advancing and often diversifying creative practice.[55] In the case of geography PhDs, it was often the geographical in particular that attracted them. So, whilst several found that a PhD in geography gave 'time for my practice,' for many it was about an explicit choice to explore and embrace what geography would bring to evolve their work. So, the ways in which these relations were figured include 'I have better geographic knowledge that I can use in developing new projects, and to help theorise my work in creative critical modes,' others described it as offering a more substantive topic based or philosophical grounding, that enabled everything from more confidence in their work, to a better understanding of how to argue for its value. Interestingly, what was common was not just particular geographical thematics, but also the questions that this posed around research methods. So as one PhD student acknowledged, 'having a better understanding of "traditional" social science methods has helped me to think more rigorously about my practice-based methods.' Whilst another suggests, 'working within a new disciplines requires me to think more explicitly about my research methods and studio work... via the geography discipline, I have had access to new ways of framing research, a different type of rigour- which I think- can work in parallel to studio work but is not transferable.' As such, in asking PhDs to contribute to art and to geography might not in fact be to ask them to do two different, separate things, but in rather to encourage a situation where they are one and the same thing.

The thesis as a site of query

If the PhD is a rite-of-passage often crucial for gaining an academic job, then it is often a site at which the forms and practices of what it means to be a researcher are reproduced, but also often extended. After all, the claims to originality and contribution to knowledge are an important part of any PhD – from application to thesis and viva. The practice-based PhD within geography is an increasingly popular form. Those who experience it as candidates, as supervisors and examiners, tend to find it to be a site of negotiation, a locus in which questions are raised about what supervision should be, how a thesis should look, what the role of writing is, and what is being judged. In many cases these points of query seem to evolve as positive points of intersection and engagement, and are what often helps contribute to the thesis's originality. As a focussed point of assessment of the quality and value of a piece of research, it is perhaps not a surprise that the thesis is a lightning rod for some of the more controversial discussions around research that emerge at the intersection of geographical research and creative practice. In particular these often come to a head in the form of the thesis itself: how should each candidate and supervisory team negotiate the relationship between creative portfolio and page, and how is this relationship made intelligible for the examiners and those who engage with the thesis at a later date. Once again creative practice-based ways of working within geography offer points for critique and reflection, and often work to extend our understandings of what geographical research is and how it is done. We owe, I think, a lot to the PhD students who often operate at the front-line of some of these debates, stepping into the midst of these debates without the ballast and confidence offered by an academic position or a secure artistic reputation, that those of us examining and supervising might have. We owe it to these candidates to ensure that the discipline that has granted them their doctorate is also able to value the generation of scholars it has created. We should ensure that our disciplinary forms and infrastructures are up to the task of the kinds of knowledge forms that these newly minted PhDs are producing. It is to an explorations of these forms and infrastructures that I now turn, considering first the pages of various books and journal articles, and then finally, the exhibition.

Notes

1 For information about Tania's work please see: El Khoury, "The Audience Dug the Graves," https://www.bac.org.uk/resources/0000/2327/Battersea_Arts_Centre_News_Jan-May_Shows_2016.pdf; https://taniaelkhoury.com/works/gardens-speak/ (accessed 12/2/2020); https://taniaelkhoury.com/2016/02/10/gardens-speak-2/ (accessed 12/2/2020).
2 Mercer, "Speculative Emblematics," see http://www.thewhitereview.org/contributor_bio/lucy-mercer/ (accessed 12/2/2020) for examples of her poetic practice.
3 https://pure.royalholloway.ac.uk/portal/en/persons/nelly-ben-hayoun(12d9e978-78ec-476a-8976-586185453442).html; https://nellyben.com/about/our-team/nelly-ben-hayoun/ (accessed 12/2/2020); Nelly Ben Hayoun, *Homo Faber and Animal Laborans*.
4 https://www.clarebooker.com (accessed 12/2/2020).
5 http://www.benmurphy.co.uk; http://www.benmurphy.co.uk/gallery#1 (accessed 12/2/2020).
6 This would include, for example also from Royal Holloway, Mia Hunt, Rupert Griffiths, Miriam Burke, Will Jamieson, Bradly Garrett. For Griffiths, Burke, and Jamieson their practices within their PhDs were natural extensions of their professional careers as artists (Griffiths and Burke) and writers (Jamieson). http://keepingshop.blogspot.com/p/about-mia-hunt.html (accessed 12/2/2020); https://frag-mentedcity.net (accessed 12/2/2020); https://geoliterary.wordpress.com (accessed 12/2/2020); http://www.miriamburke.co.uk_(accessed 12/2/2020); https://www.bradleygarrett.com (accessed 12/2/2020).
7 Candlin, "A Dual Inheritance"; Hetherington, *Issues in Art and Education*.
8 Elkins, "The New PhD"; Elkins, *Artists with PhDs*.
9 Paltridge et al., "Doctoral Writing," p. 989.
10 Loveless, "Practice in the Flesh of Theory"; Hockey, "United Kingdom Art and Design"; Macload and Chapman, "The Absenting Subject."
11 Schwarzenbach and Hackett, *Transatlantic Reflections*; Bennett et al., "Artist Academics;" Hannula et al., *Artists as Researchers*.
12 Candy, "Differences"; Haseman and Mafe, "Acquiring Know-How"; Smith and Dean, "Introduction."
13 Paltridge et al., "Two Ends of a Continuum"; Paltridge et al., "Issues and Debates."
14 Ibid.

15 Liinamaa, "Negotiating"; Macleod and Holdridge, "The Doctorate in Fine Art."
16 http://www.pride-network.eu/Texts-Tools/articleid/4/a-brief-history-of-the-doctorate (accessed 12/2/2020).
17 See for example, https://www.theclassroom.com/history-phd-degree-5257288.html (accessed 12/2/2020).
18 Ibid.; Park, *Redefining the Doctorate.*
19 Noble, *Changing Doctoral Degrees.*
20 Park, *Redefining the Doctorate.* https://www2.le.ac.uk/departments/doctoralcollege/about/external/publications/redefining-the-doctorate.pdf (accessed 12/2/2020).
21 Hockey, "United Kingdom Art and Design."
22 Park, *Redefining the Doctorate.* https://www2.le.ac.uk/departments/doctoralcollege/about/external/publications/redefining-the-doctorate.pdf (accessed 12/2/2020).
23 See for example, http://www.perditaphillips.com/portfolio/fieldworkfieldwalking-phd-2003-2006/ (accessed 12/2/2020).
24 Candlin, "A Proper Anxiety.'
25 Wisker et al., "From Supervisory Dialogues"; Manathunga, "Supervision as Mentoring"; Hemer, "Coffee as a Supervisory Technique"; Hockey, "Establishing Boundaries."
26 Quotes from a questionnaire sent to PhD students doing practice-based PhDs within geography. It was sent to 25 individuals, 22 responded.
27 Candlin, "A Proper Anxiety," p. 2.
28 Paltridge et al., "Change and Stability."
29 Rowe, "The Wordless Doctoral Dissertation."
30 Hughes, "The Poetics of Practice-Based"; Kroll, "Creative writing as research."
31 Mountz, et al., "For Slow Scholarship"; Oberhauser and Caretta, "Mentoring."
32 Interview.
33 Macleod, "The Functions of the Written Text"; Magee, "What Distinguishes"; Svenungsson, "The Writing Artist."
34 Discussed in Candlin, "A Proper Anxiety."
35 Derrida, *Of Grammatology*, p. 145
36 Derrida, *Of Grammatology*, p. 144.
37 Derrida, *Of Grammatology*, p. 281.
38 Derrida, *Of Grammatology*, p. 200.
39 Cazeaux, *Art, Research, Philosophy.*
40 http://blog.hefce.ac.uk/2017/02/06/research-takes-many-forms-this-should-shape-how-we-assess-it/ (accessed 12/2/2020).
41 Panel D overview report available from http://www.ref.ac.uk/2014/panels/paneloverviewreports/ (accessed 12/2/2020); http://www.ref.ac.uk/about/guidance/submittingresearchoutputs/ (accessed 12/2/2020).
42 Genosko, *Felix Guattari.*
43 Ibid.
44 Lesage, "Who's Afraid of Artistic Research?"
45 Ibid.
46 Phelan, *Unmarked*; O'Dell, "Displacing the Haptic."
47 Phelan, *Unmarked.*
48 Ibid.
49 Candlin, "A Proper Anxiety."
50 Corcoran et al., *ArtFutures.*
51 Mäkelä, "Knowing through Making"; Nimkulrat, "Situating Creative Artefacts."
52 Wilson, "The White Cube"; Lesage, "Who's Afraid of Artistic Research?"
53 Cazeaux, *Art, Research, Philosophy.*
54 Wilson, "The White Cube."
55 Lesage, "Who's Afraid of Artistic Research?"; Mäkelä, "Knowing through Making."

7

PAGE

Walk 'till you run out of water, the title of the downloaded article appears. 'The Walking Country', the first subtitle reads, "In this visual essay..." the piece begins, my eye moves across the text, the column of words is narrow, rather than the full breadth of the page, half of the page is taken up by a grey-scale image, a sandy S marking the trail of a snake, what looks like a GPS device for scale (Figure 7.1). The experience is heavy on texture, the grains of sand, the shaping of the form through the moving body of the snake. Without thinking my eye looks for the next column of text, it is below the image, 'Rhythm of the Walk' it reads, titling the image with a second caption (the first to the upper right of the image in smaller text details 'Figure 2, Snake trail with GPS on One Tree Point Road Broome, Western Australia. Photography by the author,' where is Figure 1 I wonder. The text below the title, shapes away from the image, narrowing off to a two-word point, almost a continuation of the sinuosity of the snake. It is surrounded though with what might be Figure 1? A line, taken for a walk across the page, I look for its origins, bottom left, '12:34:00', it reads, the single line climbs up the page in a couple of meanders, before seeming to gather pace, it doubles back, loops again, comes back around, a point, a loop, a jerky circle, a line off-left, a downward swoop, a curl and hook, a graceful curve, and then a not quite tight scribble, the line continues its jerky, loopy journey down around the tail of the text and up the side before disappearing off over the page, where I follow. Turning the page, or rather scrolling downward in my browser with the stroke of a finger on a trackpad, we re-join the text on the bottom left again, faded now, to grey it disappears under an image, 'Figure 3. Wallaby Corpse,' reappearing as a tangle of lines, still faded, it passes behind some text, still visible it crosses the two columns, passing behind more text, a density of line and loop tracks down the side of the page before disappearing off the page again. Drawn by line and form, I find I have forgotten to read. I hasten back a page, scrolling back upward to the heading 'RHYTHM OF THE WALK' and read again. It is not until the third page (p. 99), where following the looping line has taken me, rushing ahead of the text that I am

*given a sense of what this line actually is: 'Walking through spiny Spinifex grass and rocks'
the text reads 'putting pen on paper for two hours and six minutes, I end up recording 119
pages of unruly mis-rhythm (see Figure 1 [ghost image] and Figure 6)'. This 'mis-rhythm'
sets the pace for my reading of the piece, the pace and progress of that walk through the bush
running as a track throughout my experience of the paper. The starts and stops, the twists
and turns, the unruly, sometimes frenetic, at other times calm, line patterns and paces my
reading (Figure 7.1).*

Account of 'reading' Perdita Phillip's article 'Walk "Till you run out of
water," *Performance Research* 17 (2): 97–109.'

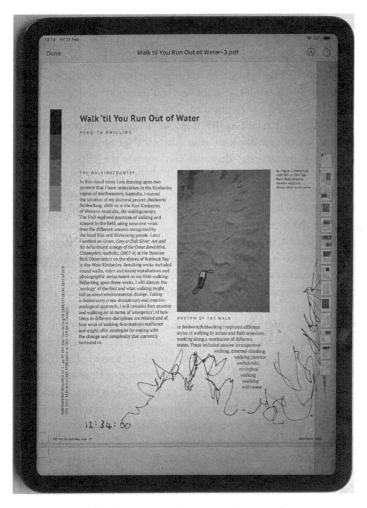

FIGURE 7.1 *Walk 'til You Run Out of Water*: page from a visual essay by Perdita Phillips published in Performance Research. (Image, Taylor and Francis)

Shaping the page

The coming together of geography and creative research practices has resulted in a rich series of outputs, from theatre, to sculpture, sound works, installation, dance, song, painting, poetry, and weaving, many of which have appeared throughout this text. Yet, despite all of this the page – of the journal article and of the book in particular – still persists and is arguably being reshaped, not least by creative writing and experimental publication practices that have intensified within geography recently. Staging a series of encounters with various pages, this chapter explores this persistence. In doing so it queries the cultures of print and publication, and of 'reading' that are being sustained and reshaped in the midst of this relationship between geographical research and creative practices. These evolving practices of the page matter, not least because as historical geographers have recognised, a 'reworking of geography's disciplinary spaces and forms of enquiry was achieved in part through a reworking of geography's print spaces.'[1] Bringing together the geographies of the book, including geographies of reading, with art histories of the audience and questions of critique, this chapter explores the production, consumption, and circulation of the pages produced through geography-art relations. In doing so, it suggests that we are witnessing a shift in what we might think of after historical geographer Robert Mayhew, as geography's grammars and dispositions towards the page.[2] As the opening encounter with artist-geographer Perdita's Phillip's journal article in *Performance Research* suggested, if pages are persisting, then as authors and audiences we might find our experiences of them somewhat reshaped. It is the questions of this reshaping that are the focus of this chapter.

Perhaps the persistence of print at the intersection of creative practices and geographical research is not surprising. After all, publication is an integral part of most academic practices, and for many of us the pressure to publish is something we became accustomed to too early in our career, and despite the diversity of creative practices, this shows no signs of abating. Critical accounts, from the histories of geographies of print-culture, to the economies and practices of publication in the neoliberal academy offer rich accounts of how it is that the production and circulation of books and articles helps to share knowledge. These accounts also understand the role of publication in economies of prestige, building reputations, helping secure jobs, grants, and so on.[3] Given these rich contexts, the persistence of print, and the continual reshaping of geography's pages – but one incidence of the 'inscription of geography' – and our disposition towards them as 'authors' and 'readers' needs further attention.[4]

The importance of the forms of geographical inscriptions to the forms of disciplinary knowledge, should perhaps come as no surprise. For, whether we look to geography or to art, there is a clear sense of how the content of knowledge and the forms in which we produce and consume it are closely linked together. For art historians shifting mediums have long brought with them shifting understandings of production and 'audience' experience. Whether this be the evolution from wall mounted sculptural reliefs to immersive environments, or the emergence of participatory art, audiences were recast from disembodied spectators to embodied sensory beings, reshaped as viewers to participants who co-produce. In the case of geography, how the discipline's knowledge is, to borrow a phrase from Innes Keighren, 'brought to book' and to print more widely, is not merely a matter of asking how and where knowledge is communicated, but rather appreciating that the where's, how's, and what's of its communication is integral to the reception of its content.[5] This chapter brings together perspectives from historical geography's diverse print-scapes, with perspectives on books and print culture evolved by art theorists and historians. Informed by historical geographers, I attempt to attend both to the micro-geographies of the book – the page layout and various devices – and to remember to 'look up… to consider the broader geographies of reading.'[6] For the geographies of the book 'takes place on every scale from that of the printed page itself to the global networks of trade and empire,' and of course the interlocked neoliberal forces of market and prestige.[7] Here, I use these historical perspectives to guide my reading of the intersection of geographical research and creative methods in the context of the growth of publishing as an artistic medium during the twentieth century.

This growth of artistic publishing – which encompasses artists' books, zines, pamphlets, experimental journals and a huge range of ephemera – was driven as much by a range of medium-based concerns, as it was by the growing critiques of the political economies of the art world.[8] This includes the print cultures that emerged to respond to the need to document temporally and spatially specific live performances, whether as a more standard catalogue of images and descriptive text, or whether as book-work in its own right.[9] As well as those forms of experimental publication that emerged explicitly as a form of critical art world practice – closely bound up with art world scrutiny around the political economies of the art world and the art market.[10] In recent years, there has been a growth in artistic infrastructure to support publishing as medium. These range from online platforms to the rise of experimental small presses, shops, and websites that specialise in dealing in these forms of artwork as well as fairs and festivals devoted to book works and artistic print mediums, whether this be high-end exclusive art fairs or those that celebrate the democratic and multiple nature of these practices.

The discussion which follows is given form through further encounters with the pages of journals, books, and reviews, created at the intersection of geography and art. Shaped by geographies of book and print, each encounter intersects scales and communities, and while recognising that practices of knowledge production, consumption, and circulation are often indistinguishable, each endeavours to draw to the fore a certain experience of producing or consuming these pages as 'authors' and 'readers.' Following the account of 'reading' Perdita Phillip's journal article which opened this chapter, the first section explores the persistence of the journal article at the intersection of geography and art, and queries how 'authors' are producing its pages as sites for critical-conceptual experiments that draw together form and content. Discussion then turns to explore the forms of geography's 'book work' that are emerging at the intersection of geography and art. It queries how, in an era in which we are increasingly mourning the demise of the monograph, the codex form is being explored and mobilised in a variety of ways. To close, the chapter shifts to explore the experiences of readers. Inspired by historical geographers who seek out the traces of readerly practices, reconstructing interpretative communities through textual evidence, the chapter begins to sketch some of the practices of reading and review that are emerging around these geographical 'pages.'

Producing the page

Geography's cultures and practices of print have long been diverse: taking in accounts of exploration published in the popular and scientific presses, the diary forms of female explorers, or ship-board accounts and albums where text and images intersect as well as, of course, the myriad printed forms of travel gazetteers and maps, and their compilations into atlases; field guides; and, more recently, monographs, journal articles and policy reports, blog posts, and so on. Whether assembled into diverse codex forms or single folded sheet pamphlets, geographers have long had lively engagements with both what goes on the page, and the very form of the page itself.[11] This includes those historical geographers interested in the shifting forms and layouts of geographic texts, their paratextual materials, and their various editions, editings, annotations, and marginalia. It also includes those geographers who have experimented with the form of the geographic text, whether that be Von Humboldt's development of an art-science style, Gunner Olsen's post-modern textual play in volumes like *Birds and Egg*, Allen Pred's experimentation with a Benjaminian fragmented form, or more recently, Sholah Kupla's artwork-essay *Hot Shots*, Tim Cresswell's poetic book-length essay on Maxwell Street in Chicago, Eric Magrane and Chris Cokinos's field guide, or Sarah De Leeuw's evolution of a thoroughly geographic poetics.[12] Indeed, the recent rise in interest in creative writing practices, has, as Miranda Ward notes, prompted the chance to explore the literature in geography in terms of creative arts.[13] Arguably, the vigorous growth of a range of geopoetic practices – from poetry to short stories, critical-creative essay experiments, as well as novels – combined with the diffusion of more attention to the conceptual work of words and style throughout geographic writing has firmly activated the page as a critical geographic space.[14] Whether it be stylistic experiments with how to write, affect, emotion, and sensation into our journal articles; a critical play with paratextual devices, such as footnotes, marginalia, and so on; or the exploratory evolution of alternative forms of handbooks, the critical potential of written geographic forms is significant.[15] This section explores some recent experiments with the page at the intersection of geographical research and creative practices, looking first at journal pages and then turning to contemporary forms of geographical 'book work.'

Experimenting with the (journal) page?

Cecilie and I sat on the floor of my sitting room, thinking about how best to represent the experiences of the urban explorative performance I had recently undertaken with her. This was the result of a workshop in which I had participated as a speaker, but also as one of the audience who went on an urban art exploration coordinated by zURBS, the urban arts collective of which Cecilie was a member. We had decided to write a collaborative piece about the urban art exploration based on our experiences of it, as the creator (Cecilie) and as a participant (Harriet). As we talked about how to try to write about this performative experience, we reflected on all the different ways we could negotiate trying to make present the experiences of this work, and decided that we should let the images and text of the day structure our discussions- so we took the map, the clues we were given and then images and decided to let these lead. At some point in our discussion, after debating for a while, these actually became the corner stones of the article, helping us structure discussions of particular incidents in the piece. As we explored the relationship between image and text, it became clear that we did not want to use images as merely 'pictures' of people 'doing' the exploration or simply 'picturing' the finds, but rather use the images in a way that tried to recreate some of our experiences of making and doing the exploration for those reading the paper. So we printed, cut, and collaged copies of the maps and images the participants had sent over of things they had collected, our own photos, and objects from my garden. We experimented with how to lay these out, with how we might create an experience of exploration on the page, at times it felt clumsy and rather heavy-handed (Figure 7.2).

FIGURE 7.2 *No More Mr Nice Guy?* In progress. (Image, author's own)

Before Cecilie and I began to discuss the form and content of our article, we were clear that we intended it for the 'cultural geographies in practice' section of *cultural geographies*. This decision was primarily driven by our sense that this was a home for the piece that would be open to anything we wanted to do in terms of experimenting with content and form. The growth in so-called 'experimental' spaces within mainstream geography journals published by large publishers has become an important site for work produced at the intersection of geographical research and creative practice. Dominated by the visual essay, but also taking in video work, accounts of sound practices, performance works, and creative writing these sections are often shorter, and have more flexible formats. Discussing why it is that they choose this *cultural geographies* section, or other similar ones in *GeoHumanities* journal (practices and curations), *ACME's* flexible format, or new sections such as *Emotion, Space, and Society's* 'creative interventions' section, interviewees mentioned it was commonly the combination of a flexibility of format combined with the ability to place one's work in a reputable journal that 'counted.' These sections offered an experimental outlet to work in ways other than 7,000 word journal articles, yet they also had the attraction of being a reputable journal – your creative experiment could still count as a journal article in a top journal. For some this was a vital way to stay true to the practice-based spirit of research whilst ensuring the all-important CV credit. Often mentioned by those practitioners new to geography was the value of being able to do 'something that feels like my stuff' whilst still reaching geographical audiences through these journals' established disciplinary readership. This accords with many of the stated aims of these sections. In the year 2000 when it renamed itself *cultural geographies* (from *Ecumene*) the 'cultural geographies in practice section' was launched, to enable, as the editorial explained:

> acknowledging, presenting and discussing how the cultural meanings of nature, environment, space and place are being engaged with intellectually and practically beyond narrowly designed academic institutions and genres. It will provide critical reflection on how practices within the artistic, civic and policy fields inform and relate to the journal's geographical concerns. As well as academic reviewers it will include; pieces from practitioners beyond the academy, from collaborations between academics and other individuals and groups. It is also open to a variety of styles of presentation, including for example, artworks, interviews and accounts of work in progress as well as discussions of cultural projects, practices and events. Of course throughout the journal it will be for our contributors to use and develop the spaces *Ecumene* opens up, almost certainly in conversations we have yet fully to imagine.[16]

Cultural geographies was not the first geographical journal to include such a section, others in which geographers were heavily involved like *History Workshop Journal* had long been committed to broader reviews of films, exhibitions, and so on.[17]

Yet *cultural geographies* was one of the earliest to actively welcome 'a variety of styles of presentation' and the spirit at work in the phrase 'in conversations we have yet to fully imagine' continues to shape discussions we have within the journal's editorial team about what might be possible. While other journals have introduced such sections, indeed, in conceiving of *GeoHumanities; A Journal of Space, Place and the Humanities*, we were clear as editors that we wanted the journal to have a 'practices and curations section,' and this was framed in terms of being 'shorter creative pieces that cross over between the academy and creative practice.'[18] As well as these 'practice' orientated sections, other journals depending on their foci have evolved sections that make space for creative practices in the context of their wider remits. *Emotion, Space and Society* for example, links their Creative/Interventions section (launched 2019) to the value of the arts and creative practice for enabling a centring of emotions 'at the heart' of submissions, 'presented in ways and from positions beyond more usual research articles.'[19] Looking across these sections, the kinds of work produced and those publishing work there, they appear to be used by those based within geography, including practitioners, as well as practitioners from other disciplines to support a range of ways that practice and research might be present for geographical audiences who might not encounter these forms of work in more art world-based contexts.

If we look more closely at 'cultural geographies in practice,' for example, one of the longer-running of these sections, we find a range of article forms. Since its launch, twenty years and eighty issues ago at the time of writing, the section has published over 135 pieces, varying between zero (on several occasions) and five pieces an issue.[20] Special edited "sections" have consisted entirely of cultural geographies in practice pieces; 'Writing geography creativity' edited by Dydia Delyser and myself in 2014 collecting together five pieces exploring these works, and in 2017, a special issue on practicing urban archives edited by geographer/ performance artist Cecilie Sachs Olsen, collecting other artists, geographers, and critical theorists in work, interviews, and others.[21] The last five years has seen a significant uptick in the number of pieces published in each issue, and looking across them a series of formats reoccur. Around fifty of the pieces sit somewhere between an arts review essay and a short journal article, exploring artworks, films, and exhibitions through more personal or creative writing forms. Another fifty or so of the pieces offer almost mini-retrospective ethnographies of creative doings. These accounts of making sometimes take interview formats, or utilise more conversational styles or photo essays, they can be co-authored with project teams and collaborating artists. They often mix reflection on the experience of making the work, and practical questions of funding, organisation, collaborative working, and so on, with conceptual arguments about what the work does for geographical topics. The remaining articles fall into the category of article as 'medium': this includes visual essays and careful assemblages of image and text (like our essay); it also includes other forms of paper where the publication, and the page itself, is clearly being explored as a conceptual space. This might be done through a series of artists pages, like those produced for cultural geographies by

FIGURE 7.3 Perdita Phillips, artist pages, cultural geographies in practice. (Image, Sage)

artist Perdita Phillips (Figure 7.3), whose article opened this chapter and how created one of the earliest examples of a cultural geographies in practice 'artists page.'[22] A further early example is offered by Annie Lovejoy, with whom I was later to produce a book-work.[23]

Another way the journal article might be considered a medium is to consider the essay as a form of document. It is into this latter category that the paper I worked on with Cecilie fits. We were trying to create for our readers an experience of the live performance that Cecilie had led and I had participated in – in a sense we wanted to document not the results, but the actual live experience on the page. This was a collective urban treasure hunt around North London that asked us to explore our surroundings and collect certain kinds of talismanic objects to narrate for the wider group. (Figure 7.4).

Following the performance art theory detailed in the previous chapter, our paper-as-document was less an act of reportage, recounting our experiences as creator and participant, and was more an attempt to recreate on the page some aspects of the experience of the performance: its ethos, for the secondary audience, who are greater in number than those who were present. If performance and live artists have long produced documentation – materials that as the previous discussion made clear are integral to the ontology of the work, rather than secondary reporting on it – then the evolution of this as a means of article making is important. This is important not just for enabling the participation of those who

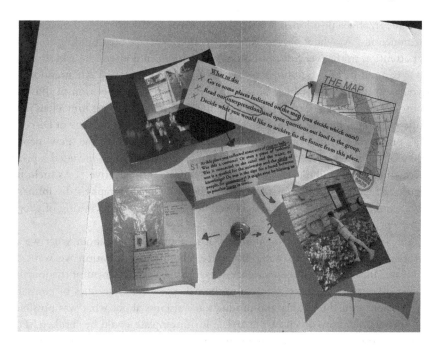

FIGURE 7.4 *No More Mr Nice Guy?* In progress. (Image, author's own)

create performance and live art in publication practices integral for their ongoing role in the academy, but is also an important way to reach secondary audiences, and specifically geographical audiences many of whom are less likely to be aware of, or present at, art world events.

Producing the paper with Cecilie felt like conducting a series of experiments with the composition as a conceptual and experiential thing, rather than focussing on content, with questions like layout a backgrounded automatic element aside from the odd <inset image in here> call out. Cecilie and I sat on the floor surrounded by a range of materials and tried to create compositions of objects, image, and text. A range of textual, spatial, and paratextual devices became important in evolving a discussion of the page as a conceptual place, in each case aesthetic-conceptual concerns negotiated academic conventions. For example, an important element of this text was the relationship between image and text, undercutting the primacy of the text being crucial to the work. Images were less illustrative of points made within the text, but as well as carriers of information were part of a complete compositional unit. While our attention to the layout and spaces of the page, and those of other papers within the 'cultural geographies in practice' section might seem somewhat incidental we should not under-estimate the importance of the page and its layout in thinking through the 'inscription of geography.'[24]

As Robert Mayhew demonstrates looking at historical geography texts, attending to geography's inscriptions can be a key determinant in understanding how geography is positioned as a scholarly enterprise.[25] Deploying a 'materialist hermeneutics,' Mayhew moves from paper, ink, typeface, and binding to draw our attention to the 'very disposition of space itself'; in doing so he asks us what these can tell us of geography's spaces and forms of enquiry?[26] An attention to layout and margins; to tables, boxes, and sequences; to the position of illustrations, maps, and drawings and their form in relation to text offers him a lens into shifts within geography. As he explores, this was a shift from the Age of Discovery's print space of description, reference and the gazetteer, to the 'new' geography of the nineteenth century and its Imperial vision which included a print space cultivating legibility, argument, explanation, and narrative.[27] As such, he argues we can grasp a sense of how a 'reworking of geography's disciplinary spaces and forms of enquiry was achieved in part through a reworking of geography's print spaces.'[28]

Yet, in producing pages for journals, it is not always just about what we as authors might want to do with that space, about the disposition we want to cultivate. Indeed, my own experiences of working with the form of the journal article, like many accounts offered by those I spoke to, were ones of constant negotiation with the editors and production assistants about what was possible: in other words, how much of the article form template could be 'broken.' For some of those I spoke to, this was an engaging process, posing a set of 'generative constraints'; for others it was a challenging and frustrating experience. In many cases content-based issues were solved through innovative use of paratextual devices, such as abstracts, footnotes, and references. The abstract, for example, has evolved less as summary of the article's key points and more as a site of demystification, either of the production of the piece that follows or as a space to spell out the link between form and content. Ian Cook, for example, in a 'cultural geographies in practice' piece offers the article's biography: 'the process began as an emailed invitation, moved online to blogged conversations, and produced a paper that aimed to stretch the journal format to capture and continue that conversation.'[29] While, Cecilie and I positioned the abstract as programme notes for the 'performance' which followed, including detailing our various roles, 'Harriet observes from a participant's perspective, whereas Cecilie discusses it as a facilitator of the workshop.'[30] Whilst much can be done in these paratextual sites, there are of course, limits too.

Returning to the encounter with Phillips's pages that opened this chapter, her on-line marginalia offers us fascinating insights into the publication process. Phillips has usefully collected online her work, writings, and artistic practice produced before, during and after her PhD, which involved a visit to Goldsmiths London and Royal Holloway Geography, where she took elements of the MA in Cultural Geography. Alongside the links to the pieces are a series of marginalia comments on each one, often including reflections on the publication process for each piece.[31] Alongside 'walk till you run out of water,' she writes 'an image essay on walking, environmental change and resilience beautifully published in performance research: a journal of performing arts.'[32] By contrast, is the down-loadable file of the Australian Geographer paper, *Doing art and doing cultural geography: The* fieldwork/fieldwalking *project* (2004).[33] The file contains the original layout Phillips proposed for the paper together with the final journal version. She does not comment in detail on the differences, but does note that 'during the publication process low quality jpegs were substituted for the original illustrations which are re-introduced here. Note: the word fieldwalking has also been altered to field walking during the editorial process.'[34] The original proposed (but not produced) layout cut together images and text, jigsawing them around images, rather than in the usual columnar format of the page. Some of the pages used full-bleed black backgrounds onto which white text was superimposed. Images carried over several pages, text boxes floated slightly transparent over full page images. Text, image, layouts, all failed to hold still and all failed to subscribe to the standard journal format. On the first image of her proposed layout a diagonal text box has been overlain, like a sticker or a stamp: 'This should be formatted as normal for the journal.'

Not only do problems with these innovative forms of text occur pre-publication but also afterward too. On going to look for Phillip's early artist's pages in *cultural geography*, a two-page spread (Figure 7.3) I struggled to find it. I had first encountered the work in a paper copy of the journal, and had I not been able to see the reference to it on Phillip's website, I might have thought had dreamt it. It was an electronic ghost piece, not locatable in any electronic search of the journal's archive. I eventually encountered it again when going through the back catalogue of 'cultural geographies in practice' pieces. Lacking an abstract and being so short it had, in the online indexing of journal content, been folded into another piece prior to it. This was corrected when I pointed it out, but it speaks to the ways in which these kinds of works – short in form, not as easy to understand as an article as two full pages of images – pose challenges to standard journal conventions.

As even a cursory look at 'cultural geographies in practice 'and other sections demonstrates, geographers are becoming disposed towards the space of the page differently. Rather than just a carrier for content, to be arranged according to standard formulations, the page is emerging as a conceptual space in its own right. If Mayhew could write of an emergence of a new 'geographical grammar,' in which shared forms come to mean something (the shared geographical print space and grammar of the Age of Discovery),[35] I think a case can be made, if not for a fixed grammar of these contemporary interactions, then certainly for a heightened awareness that a geographical grammar exists that we might want to use otherwise. In other words, we could suggest the cultivation of a disposition towards the geographical page that is increasingly aware of how we might use it differently. Clearly these are decisions to be reflected on in ways that align with our own research topic and needs, but as the example of Perdita's pages makes clear, these are not decisions over which we, as authors, have final and complete control.

Geographical book-works

The book was small, square, four inches by four inches, with a black card cover, with a fold-out end paper. The stock we chose to print on was recycled paper of a high quality, a slightly cream colour with a good texture. It was a small book, just fifty odd pages, with images and some text about knowing place. The entries used the space of the page as a composition space, experimenting with how different forms of layout could help us evolve and reflect on different ideas of place. We bought together hand-drawn sketches, photographs, diagrams, and maps, with quotes from local people, from oral historians, and from academics (Figure 7.5).

The book was designed to accompany a series of sessions on the geographies of art, held at the 2009 Royal Geographical Society conference in Manchester, the theme of the conference was the geographies of knowledge. In discussion with the artist Annie Lovejoy I was working with at the time, we conceived of the artists' book – 'Insites'. We decided on a limited run of 1,000. A cross between a notebook and a set of reflections on geographical knowledge, we placed these books in the conference bags handed to conference goers, where they sat along with the leaflets and publicity materials, many we suspect got thrown out. Late on in the process, after talking to our colleagues about our plans and receiving some blank looks, we created a transparency insert in the book to highlight the making process and to talk about the book as a form of geographical knowledge.

Despite debates about the book, and its future as a geographical research output, especially in the form of a monograph, book-work at the intersection of geography and creative practice has been expanding of late.[36] For some the 'contemporary

FIGURE 7.5 *Insites:* An artists' book, Annie Lovejoy and Harriet Hawkins (2009). (Image, author's own)

merit evaluation culture' challenges the time needed to research and write scholarly monographs; the monograph is increasingly seen as 'outre, an indulgence or both.'[37] For many within the humanities, however, the scholarly monograph remains a *raison d'être*, despite the challenges and risks. Interestingly too, alongside the morphing of the journal pages, comes a diverse sense of geography's bookwork, including experiments with the artist book form like that which opened this section (Figure 7.6).[38] What follows considers the perspectives offered on geography's book work through the forms of publication experiments.

Experiments with books have long been a part of geographical practices. Whether we think about Allen Pred, or Gunnar Olsson, both of who pushed at the edges of the monograph form with creative-conceptual experiments in text and form.[39] Recent inheritors of this tradition might include Tim Cresswell whose book Maxwell Street combines an experiment in place-writing with histories of a sense of place.[40] Alongside these experimental monographs, are those for whom the codex form offers other opportunities, whether that been in the form of poetry books, short story collections or novels, as well as photo-books.[41] We might also reflect on the iterations of the form 'edited' collection that have emerged. Edited collection feels like a poor description of these experimental forms but it covers the nature of the multiple authorship implied. I am thinking here of texts such as *Patterned Ground* and *City A-Z*, both of which experimented with how a multiple-entry volume might do fascinating and engaging things.[42] A wonderful contribution to this recent field is geopoet Eric Magrane's

FIGURE 7.6 A range of different geographical book-works. (Image, author's own)

co-production with Chis Cokinos of *The Sonorean Desert: A Field Guide.*[43] Here the form of the field guide offers an inspiration and departure point for a critical creative text that draws together local, national, and international poets, writers, and illustrators, to offer a guide to place and environment. Its form ensures it is engaging in both the armchair and the field. Others have used the codex form differently to produce a series of artists' books and small press forms such as the poetry chap book, or the artists' multiple (Figure 8.6).[44] These volumes, produced either alone or through a small press (often an independent publisher producing very small runs of limited books and magazines), take a range of forms. Often placing as much focus on the aesthetics of the volume and the choice of paper, as well as the written word and image form, these volumes take the book as a medium. As exciting as the breadth of geography's book work is, I want in this section to explore how these evolving codex forms also open questions of the negotiation of different forms of participation within academic geography's cultures of print and publication, including the politics of circulation.

A politics of circulation

Artist's books are a tricky form. Indeed, for some the only way to define them is by their ambiguity, whilst others wrestle with various ways of trying to capture them. From the art historian Lucy Lippard for example, 'artists' books are not books about art or on artists, but books as art.'[45] One of the reasons artists' books have become so hard to define has been that a denial of organisation, integration, and definition is part of their ontology. Indeed, Joanne Drucker's attempts to build a critical framework for book art practices were widely denigrated as a betrayal of the very ethos of the artists' book.[46] These books can be all words, all images, or combinations therefor. At best they are 'a lively hybrid of exhibition, narrative, and object – cinematic potential co-existing with double-spread stasis.'[47] If some artist book-makers experiment with entwining form and content as part of the book's conceptual work, others foreground critical reflection on the form of the book itself, its materiality, its production, and its economies of consumption and circulation.[48] The histories of the artists' book are often entwined with the evolution of art in the mid-twentieth century and the critical standpoint artists were developing on the economies of art and its gallery and market-based existence.[49] The artists' book emerged as a form as a result of its denial of the forms and composition of the art world and especially the art market. For many it has been understood as a deliberate practice to destabilise and operate through and outside of the art world canonical traditions – long symbolised by large paintings, rarefied value systems, singular objects, and expensive material forms – including, of course, gilt frames and large galleries.[50] By contrast the artists' book has been held up to be portable, durable, inexpensive, intimate, non-precious, and highly replicable as a format.[51] For some the promise of the artists' book was in its critique of the arts market, for others it was in the ways it offered an accessible format easily distributed, effective for both struggling artist as well as reaching wider, less rarefied audiences.[52] As Lippard notes, for some, 'artists' books mark a genuine historical moment of dissatisfaction with art's outreach, and declaration of independence by artists who speak, publish and at least try to distribute themselves, bypassing the system.'[53] Whilst there are, of course, other histories of the artists' book work we can tell, this emphasis on the politics of circulation and economies of production strikes a similar note with some of geography's alternative book forms.

There has been a growth in recent years of geographers creating artists' books and also turning to small-press independent publishing houses to produce books with high-quality production and aesthetic values at an accessible price.[54] A core example from the UK is offered by the geographers who have published with the small press Uniform Books. Uniform Books, stocked in bookshops and galleries, and often found at book fairs, bills itself as an, 'Independent imprint for visual and literary arts, cultural geography and history, music and bibliographic studies.'[55] Based in Axminster, South West England, it has a growing catalogue of books by cultural and historical geographers, as well as related titles on landscape and environmental topics. Indeed the press might be thought of as publishing some of the most interesting GeoHumanities texts in recent years.[56] Four recent offerings (Figure 7.7) detail the interest of geographers in this space: *Anticipatory Histories,*

FIGURE 7.7 Some recent Uniform Books. (Image, author's own)

an edited collection of short pieces, evolved through a network that sought a lexicon for current environmental challenges.[57] *Modern Futures* is an edited collection by Ruth Craggs and Hannah Neate containing essays that explore modern architecture. *The Regional book*, by David Matless, offers forty-four examples of place-writing that form descriptions of various locations on the Boards, a wetland region of eastern England. While *Visible Mending* is as a collaboration between geographers Catilin De Silvey, James Ryan, and photographer Stephen Bond.[58] These volumes privilege high production values, and are affordably priced, costing around £15 each. It is often the combination of aesthetics with the wider audience that authors cite when reflecting on their choice of this press.

Following historical geographers and taking the 'review' as a site for understanding the economies of publishing, it seems clear these are books that reach and delight a wide audience.[59] They are celebrated in geography journals, online websites, creative practice journals, newspapers, environmental journals, online critical literary sites and blogs alike. Further, they are available to buy at interesting locations – making their way into art galleries and institutional book shops. These include, The Royal Institute of British Architects, Tate Modern and Britain, the Institute of Contemporary Art, as well as trendy and established book sellers, such as Daunt Books and the London Review of Books, and into numerous art and print fairs as well as book fairs. As such, these small press publications offer an attractive chance not only to produce a beautiful book, but also to bring that work to a different kind of audience.

Whilst small press book fairs and artists' book shops might not be normal or easy spaces for geographers as a discipline to access, there are, perhaps unsurprisingly, a series of online sites that combine the production and distribution of artists' book-works. One platform that some geographers have used is the website diffusion.org.uk.[60] Diffusion.org.uk was developed in collaboration with the artists group Proboscis. London-based Proboscis works through an ethics based on collaboration and what they call 'co-discovery.' Their practice is grounded by a desire to communicate across boundaries, whether these are between different partners and communities, or within research contexts, say for example between geographers and anthropologists.[61] Part of this ethos has involved the evolution of new types of open-access publication practices. Their interface of 'hybrid digital/physical publications' enables those with time and internet access to access their free digital library of book-works to be explored online, or to be downloaded, printed, and folded. The online interface also enables the user to make their own book work and add it into the library.[62] The geographers currently found in the library include Hayden Lorimer, in collaboration with Kate Foster (Crossbills); Kathryn Yusoff (LANDSCAPE: 3ACTS Return/ Circulation/Dispersal) and others such as economic geographer Andy Pratt, who worked with Proboscis on questions of the digital city.[63]

Closely, although for some contentiously, related to the artist's book, is the form of the zine. The 'zine' – short for magazine – takes some of the critical production values and political economies of circulation that are raised with the artists' book to their perhaps extreme end.[64] Zines, often associated with the feminist and punk sciences, as well as underground politics, are cheap, easy to produce, often cut and paste forms of publication.[65] Their aesthetic is often that of the hand-made, privileging older forms of media- such as type writers and printing presses, before being run through photocopies, and then stapled or stuck together, cheaply producing copies, sometimes many thousands, for distribution.[66]

Zines seem to have been having somewhat of a moment within geography of late. They have become used in teaching and in particular within participatory research where their grass-roots nature, DIY format and materiality is suited to the ethos of such ways of working.[67] Jen Bagelman (a geographer) worked with her sister Carley Bagelman (an artist) to use zines as part of a programme of re-search and teaching around food stories, and food justice.[68] In their discussion of this work in a paper for ACME they explicitly relate the economies and practices of zoning to a critique of the neoliberal academy.[69] Elsewhere Bagelman dis-cusses her use of a recipe book format as part of her participatory work exploring migrant experiences of sanctuary cities.[70] Other academics value the zine as a poly-vocal format – responsive to the challenge of field-based working across multiple sites and with many communities.[71] For the 'Camps to Cities' team, for example, a fifty-page printed zine, together with a photographic-based website, offers their project focussed on European migration and citizenship the means to 'think across' their field sites – Athens, Budapest, Berlin, and Paris (Figure 7.8).

FIGURE 7.8 A page from the *Camps 2 Cities* zine. (Image, Camps2cities)

It also offers them a means to 'think with' their research participants as well as scholars from different disciplines.[72] Arranged on the page, the volume develops stories, voices, and reflections that explore the experiences of migrants. As they describe it, the zine's multiple registers of material collaged around four themes: waiting, uncertainty, dwelling, and solidarity. It 'marks moments that shaped the back stories of refugee and migration politics through this past year, and aims to open up questions for future research, activism and more humane policies.'[73]

If for many geographers alternative forms of book-work reach diverse audiences in engaging ways, their highly circulatory and often ephemeral form also poses challenges around how to locate, study, and ultimately archive these forms. Sarah Hall's participatory work on lived experiences of austerity in northern England is an interesting example.[74] During the project, she worked with a zine maker, and while the project's academic papers make little mention of the zine, it is heavily featured in the project's wider presence, on webpages, on Twitter, and in the University's promotional material. This includes videos about the zine, its making and the accompanying exhibition, as well as a downloadable copy of it.[75] Not all zines have such records however, and in many cases projects can be overlooked or go missing. This is a shame as while the academic context may not be their primarily focus, their presence and value is interesting, not least in thinking about the critical economies of the pages that exist at the intersection of geographical research and creative practice.

Challenging pages? "Reading" at the intersection of geography and art

> I find it very hard to intervene in texts that are written in poetic form like that. With a traditional social science structure there are a series of points, making it easier to intervene (if not so enjoyable to read or listen to!). When you have a poetic expression, one's engagement is different. I think 'I really enjoyed that', but the poetry of some current landscape writing can make it very hard to intervene in the text. Articles can be almost hermetically sealed, beautifully written stories, but how do you intervene? Do you intervene aesthetically? Or do you intervene in another way?[76]

Tim Cresswell's comments, in a written version of spoken conference remarks for a session 'Landscape, Mobility, Practice,' at the Royal Geographical Society Conference in 2006, begin to raise some of the questions these reconfigured pages and book forms pose for geographical readers. These questions include, how do we 'read' these texts, and by extension how do we review them, either within the peer-review process or in published reviews, and furthermore, how do we put these 'texts' to work within our own scholarship? With the emergence and evolution of new forms of creative practice come questions around the kinds of audiences created and the modes of critique demanded. Claire Bishop, for example, has written on the emergence of the embodied sensory audience for installation art, in contrast to an apparently disembodied 'viewer' of painting.[77] Whilst Jen Harvey, Bishop, and others have observed the emergence of the audience as participant and even co-creator in relational and participatory art.[78] As earlier chapters have indicated, this shift in the nature of art and so the ideas of audience is indistinguishable from the questions of the form art critique needs to take. To close this series of encounters with the pages created at the intersection of geographical and artistic practices, I want to turn to these questions of audience and critique through an exploration of some emerging geographies of 'reading' required by these pages. I put reading in quotation marks here because such practices are often those of consuming images too. Discussion will explore how reading practices reveal and potentially challenge backgrounded disciplinary habits of knowledge reception and critique.

The spaces, practices, and materialities of reading are a relatively new area of preoccupation in the wider attention to the geographies of the book. Yet, as historical geographers note, conceptual and methodological challenges circulate around the

> need to defamiliarise all too familiar [reading] practices in order to revel how it has fitted into and shaped the processes by which ideas are made and moved, and the ways in which identities are made as people are moved in different ways by their reading.[79]

The details of these practices are important to attend to for, as Livingstone and Withers note, 'beyond the area of the history of science- where notions of comparability, standardisation, credibility and fact are particularly highly charged- how readers read books might be quite different.'[80] What comes to be 'highly charged' in the context of creative practice and geographical research will be the focus of this discussion. Important, for, as Barnett's studies of geographical text books demonstrate content, form and paratextual devices – from titles to typeface and cover designs – can come to constitute new reading publics.[81]

My exploration of some of the geographies of reading pages created at the intersection of geography and art has been inspired by the forms of archival work and practices of tracking readerly traces that historical geographers of the book have evolved. More intrepid than I, historical geographers have sought the habits and dispositions of their long-dead readers through detailed attention to prefaces, translations, footnotes, illustrations, and omissions, tracking the evolution of papers through reviewers and editors comments, where marginalia spaces can become conversational sites.[82] For my part, I went looking on- and off-line for the 'interpretive communities' of readers. Together with my own experiences as author, editor, reviewer, and reader, I focus my discussions here through the figure of the reviewer. This includes both the publicly published reviewers, and the imagined peer-reviewer summoned up by review instructions for a range of journals that publish practice-based research. The very existence of publicly available reviews for texts at the intersection of creative practice and geographical research – on blogs, in society newsletters, and in literary magazines – evidences a different kind of reading and reviewing public than that we might expect for a standard academic monograph. What is clear however, reading across the myriad forms of review is that this is often a reading public in active constitution. In other words, the reviews often openly frame questions that the interdisciplinary nature of the texts raise, thus ensuring that some questions of the relations between these forms of knowledge, their histories as well as questions of expertise and evaluation become embedded in the circulation and consumption of these kinds of texts.

The peer-reviewer

I find reviewing and editing at the intersection of geography and creative practice hard. I spend much time worrying whether I am qualified to peer-review this work. I often find myself dealing with things neither within my area of practice expertise (I have no real skills to review creative writing for example), nor in my field of substantive expertise. While I am very happy to be a catch-all reviewer for many forms of creative practice-based submissions, I always find myself qualifying my review for editors and for author. This usually involves my detailing my relation to the practice and topic, and heavily caveating my reflections or posing them as questions, which perhaps makes for a somewhat frustrating review for editor and author alike. I often feel like these reviews end up forming an active set of reflections on the terms of peer-review and academic judgement itself, and where I feel these criteria and my own abilities fall short in the face of the work I am being asked to pass judgement on. My reviews often end up feeling more like mini essays in and of themselves, whether because they offer almost reviews of the review process, or whether they end up feeling more like a response to the piece in terms of how it inspired me, how it generated new thoughts, or moved and excited me. Whilst these elements are present within my reviews of more 'standard' 8,000 word academic papers, they feel less dominant and more of an aside or maybe the introduction, whereas with creative work they are often the central thrust of my reflections. While I do use more standard review forms, it often feels churlish and against the spirit of a creative piece to meet its generative inspirational and emotional force by pointing to gaps in literature, requesting more theoretical nuance or details of methods; furthermore such requests would often be thoroughly inappropriate. I also have to resist focussing my reviewing attentions on the expositionary text that often accompanies the creative material, or only responding to the creative material by way of reflections on its relation to that text (Figure 7.9).

It is interesting to situate my experiences in the context of the instructions to reviewers offered by journals which specialise in practice-based research, and the debates about peer-review that these journals often openly discuss. The Journal of Artistic Research (JAR), whose instructions for reviewers are quoted over the page, is an 'international, online, Open Access and peer-reviewed journal that

(a) Journal for Artistic Research
Peer Review Form

Dear reviewer,

Before you start working on the submission and your assessment, please take a look at the various sections of this form and structure your review accordingly.

There are 10 sections:

1. Reviewer self-assessment
2. Interest and relevance
3. Potential
4. Exposition of practice as research
5. Design and navigation
6. Ethical and legal concerns
7. Conclusions and revisions
8. Recommendation
9. Confidential message to the Editorial Board
10. Feedback

The questions that are asked in these sections should be considered as a guideline for your review. Please feel free to deviate from them should you think the submission of your review requires that.

(b) 4. How does the submission expose practice as research?

JAR is open to submissions from various methodological backgrounds, as long as they expose practice as research. By this we mean that the submission exposes, translates, stages, performs etc. the practice it presents so as to engage with its own meaning, to challenge existing epistemic horizons or to offer new insights.

Please take into account:

- Whether or not the submission contains a description of the question, issue or problem that is explored, and if not, if such an omission matters;
- Whether or not the submission shows evidence of innovation in content, form or technique in relation to a genre of practice, and if not, if such an omission matters;
- Whether or not the submission is contextualized and the context is referenced, which may include social, artistic and/or theoretical issues, and if not, if such an omission matters;
- Whether or not the submission provides new (kinds of) knowledge, interpretation, insights or experiences, and if not, if such an omission matters;
- Whether or not the submission's methodology if adequate and thorough, and if not, if such an omission matters.

Ultimately, a submission may successfully expose practice as research despite disappointing review. If applicable, please state conventional academic criteria for the assessment of research. If applicable, please state where the breaching of such criteria is detrimental to the submission.

FIGURE 7.9 Extracts from the *Journal of Artistic Research* peer-reviewer instructions.[83] (Image, author's own)

disseminates artistic research from all disciplines.'[84] Founded in 2010 JAR now publishes three issues a year; it is currently on issue 19. Importantly, the JAR format for publishing artistic research is the exposition, which requires authors to 'combine text, image, film, and audio material on expandable web pages, challenging the dominance of writing in traditional academic research.' Rather than a manuscript submission system, JAR is based on a 'Research Catalogue,' which involves authors, creating, designing and submitting an exposition.' The 'research catalogue' is 'a searchable, documentary database of artistic research work and its exposition. The Research Catalogue is an inclusive, open-ended, bottom-up research tool that supports the journal's academic contributions.'[85] This catalogue is built in a 'rich media web format' and as such cannot be downloaded in PDF form. You can self-publish within the research catalogue without necessarily submitting an exposition for peer review.

JAR has been, and continues to be, an important part of the practice-based research landscape, not least for its editorial discussions of the challenging of publishing practice as research. As they note, 'in JAR artistic research is viewed as a developing field where research and art are positioned as mutually influential. Recognising that the field is ever developing and expanding, JAR remains open to continued re-articulations of its publishing criteria.'[86] Peer-review, and debates about what form that should take for practice-based work, is a crucial part of these ongoing re-articulations. In his opening editorial Michael Schweab noted,

> the publication of JAR 1 is a particularly important occasion for us, because, over the past months we have been looking at the key problem of how best to peer-review artistic research. We have come up with a solution we are tentatively proud of, aware that the peer-reviewing we propose is vital for the journal's credibility, artistic or otherwise. We are interested to hear what you think.[87]

The editorial for issue nineteen, revisits the issue nine years after the journal's founding. It notes JAR, 'from the very outset has been sceptical about criteria that may be applied during the review process.' What emerges in their almost of a decade of publishing and reflecting on peer-review is a call for what they term 'epistemic solidarity.' They ask peer-reviews to depart from their role as experts, less 'geared towards actions of groups of people or shared identifications with a higher principles' rather, epistemic solidarity is 'more of a local, ephemeral and imaginary process by which one peer, practices empathy … in such a way that potentials that become tangible, that are not rooted or at least not fully rooted in his or her expert knowledge.' Reading such reviews they describe as an experience of 'reading sections of an anonymous letter to a friend.' This sense of solidarity between author and reviewers, as JAR editors suggest, an 'appreciation of another peer's struggle for understanding,' is something I recognise as an editor, author, and reviewer, although most of us know that peer review is sadly not always like this. There also seems to be something important here in the way reviewers are being asked to respond. JAR reviewers are directed towards a series of criteria the submission must fulfil – a question, an innovation in practice, a context, a contribution to knowledge, a clear method. These are akin to those check-lists that now form part of many geography journal's electronic peer-review systems. Yet important for JAR is the caveat given on each of these criteria – 'if such an omission matters.'[88] This all-important caveat means that the reviewer is being asked to judge the piece highly reflexively and is being asked to indicate whether the scaffolds of a research article usefully apply here. They are being called, in short, to apply their own judgement to the relevance of the standard elements of the peer review process. It is clear that, as the editors write, this is an emerging field that requires ongoing articulation and one whose form is not yet clear. This is a tough ask, in order to do this well you need to meet the piece on its terms, you need to inhabit it in a certain way to ensure you understand what the right terms upon which to assess it would be. To simply apply blanket criteria would be to do the piece a disservice and to misunderstand what JAR desire from their peer-reviewers. Of course, good peer review, of creative practice or not, could be considered akin to this, but we are perhaps less aware of it as a formalised part of the process.

Those asking questions around what artistic research means for art criticism often discuss a bifurcation between forms of response. One the one hand are those responses that contemplate the aesthetic form of the presentation of the research objects, whilst on the other are those modes of engagement that explore 'the artist's frequently idiosyncratic research endeavours in all their aspects and details.'[89] In other words, do you query the aesthetics of the work, the form, the presentation, the experience you and other audiences might have, or do you query the 'research contexts and methods implied by and evolved by the work, and relate them to epistemological purposes.' These are, as Holert and others make clear, very different modes of perceiving the artistic object: modes of contemplation that rely on either theoretical judgement or aesthetic judgement.[90] Yet perhaps such bifurcated approaches that suggests there is a choice to make when consuming – as reviewers, as audiences – these forms of artistic research is unhelpful. Indeed, it seems rather as if what is needed is less to choose between the two approaches, as to work out how to reconcile them, which brings its own epistemological challenges. Indeed, I can think of little art criticism that would ever thoroughly delink the two perspectives. Indeed, most art theory and history with any currency seems to fundamentally link theory and aesthetic reflections. For peer-reviewers the question seems to be how to respond, especially when you might, and your discipline more generally might, lack the literacy needed to do so. Interestingly, we see in published reviews a kind of dramatisation of these discussions, as commentary goes back and forth with some of these questions. It is to these reviews I now turn.

The published review

> Film stills with photographs which often occupy as much space as the text, and novel graphical design, *Modern Futures* has a vibrant visual aesthetic. Both the publisher and editors should be commended.[91]

This is a book for looking into, within and across its various contributions, to unsettle self-evidence truths, to problematize the easy configurations of naturalised histories and to suggest other stories from the archive.[92]

We rarely read alone. Indeed, book scholars have numerous appellations for our readerly communities, from 'invisible colleges' to 'interpretative communities.'[93] What is clear is the 'inescapably collective character of interpretation' and thus how any individual reading should be situated in the context of 'a reader's membership of a community sharing some foundational assertions and interpretative traits.'[94] Historical accounts of review cultures reinforce this idea. As Nicolaas Rupke writes of the reception of early GeoHumanities scholar, Von Humbolt's texts, for example, 'reviewers stage mange texts and steer readers in particular directions.'[95] Here I want to explore the review culture that has grown up around the volumes published by Uniform Books, the small-press discussed earlier. I am interested to query the terms of judgement being used, the type of critique evolved, and to consider how these reviews, like the disposition towards peer-review promoted by JAR, often call the reader towards the terms and forms of critique itself, requiring that we actively reflect on what we bring to these texts.

The cultures of review emerging at the intersection of geographical research and creative practice are rich and as many report challenging. Book review sections in major journals now increasingly include exhibitions, films, festivals, and experimental texts alongside reviews of academic monographs and edited collections. Reviews of books by Uniform Books are not only found in academic journals, but also in literary magazines, online art and literature sites, specialist environmental blogs, local literary society magazines, news websites, local history societies, the list could go on. Taking the temperature of these reviews tells us quite a bit about their feelings about the content and form of these works. Academic reviews of experimental book works, almost always note the difference in writing style and the use of images and design. Indeed, as Stephen Daniels observes of Bipolar, 'the volume is as much material, as intellectual, artistic as academic. *Bipolar* is a fine and imaginative example of bookmaking as well as a culturally provocative text.'[96] Some academic reviews are more attentive than others to the nature of different contributions, 'short essays, reports, testimonies, diaries, histories, graphs documentary photos, and artworks.' Other reviewers overlook the conceptual force of design labour and image all together, merely noting volumes to be 'lavishly illustrated,' as if this was a luxury rather than a fundamental aspect of content. Another set of reviews manages to equate rich illustration with public engagement and a sense of readability. For some, what is 'slim but readable,' for others an important signal of something different: 'care and enjoyment is also evidenced in the material form of the book.'[97]

One reoccurring element of many academic reviews of experimental book-works found in geographical journals is a dramatisation of the struggles with the forms of criticality these texts evidence. In the case of reviews written by geographers for geographical journals, there seems to be a tendency to make a, perhaps, false comparison with journal articles. Here, praise is given to public accessibility but lost for a 'lack of criticality.' There are at times revealing critical asides: 'the design of the book allows for a playful exchange between text and image (perhaps too playful...).'[98] In one case for example, after being praised for brevity and accessibility, a note of caution is sounded:

> The book also, however, signals some of the limitations to this kind of publication, in that the desire to produce something that is accessible to public as well as academic audience may mediate against the kind of sustained critical engagement that would be expected in a peer-reviewed journal article.[99]

It is common to find these forms of work being reviewed as coming up short in the context of academic expectations. This is signalled by phrases such as 'lightly referenced' or 'raises more questions than it can answer'; indeed the productivity of raising questions might be the point and perhaps the delight of the text.[100] Oftentimes, geographical reviews give less attention to layout and image, or the 'aesthetics' of the text, and the conceptual force of these elements. This is made clear when you read the exception. Tim Edensor, for example, in his review of *Visible Mending*, a collaboration between geographers and a photographer produced by Uniformbooks, honours the form of the book as a sequence of 'photographs and brief observational vignettes' (although he notes there is an essay at the end).[101] His review attends to the photographs making them as central to his discussion as they are to the book as a whole.

Even more challenging than visual content seems to be experimental textual form. Where this tends to be engaged is often in terms of issues of literacy and legibility, rather than noting it as a conceptual device. John Pickles wrote in 2007 of Gunnar Olsson's (1980) *Birds-in-Eggs/Eggs in Birds*, 'flipping of thought and printed page as it probed repeatedly the limits of language and thought until the reader felt it was his head hitting against that ceiling.'[102] More recently of Cresswell's experimental style in *Maxwell Street*, 'there are moments where I would have welcomed a linear narrative, less tangents... it is fairly experimental in its presentation, and the effect is mostly enjoyable.'[103] What emerges from these reviews seems to be a community who brings to the texts a tentative embrace of their novel (for geography) aspects, but a community that is very much still conditioned by sets of expectations around what texts should do and be: critical, in-depth, offering clear answers. There is a valorisation of public accessibility which is what these texts are often understood as, but the embrace of their creative forms on their own terms is less than wholehearted ('perhaps too playful ...' 'mostly enjoyable'). Contrasting those reviews which have hailed from geography, with those produced from within more creative practice-based contexts and there is a striking difference in attention to the aesthetic form and stylistic effect of the work, and also the ways that conceptual challenges of grappling with form are approached.

One key difference often found between geography-based review cultures and those of other contexts is the degree to which other cultures of review attend to cultural intertextuality. In other words, whereas the academic geography review might situate the volume in similar academic literature, other reviews attend to creative inspirations and so on. For example,

> this Uniform book is not only elegantly designed, but is uniformly erudite and well-written. The spare, enigmatic prose style — something of a house style for its publisher- the small but perfectly formed Uniform Books- reminded me of one of John Cower Pwyes's lesser known works.[104]

Reviews such as this offer a useful reminder that when we think of creative practice based work, we might find value in thinking about the complex assemblage of other texts, and of course practices, that are its inspiration. Gillean Beer's account of how Charles Darwin's wide ranging and voracious reading habits fundamentally shaped the nature of his scientific argument, demonstrates how books 'do not stay inside their covers. Once in the head they mingle. The miscegenation of texts is a powerful and uncontrollable force.'[105] Further, as readers we don't just bring to bear conceptual geographies but also other forms too. As Livingstone observes, 'readers come to texts with different reading histories, and read in light of their literary genealogies.'[106] In a similar vein creative works don't stay within their exhibitions or gallery spaces, or poetry books, but instead mingle in new works, and its valuable to recognise this when reviewing them.

Another core aspect of reviews from beyond geography is their reflective approach to the style and form of the writing. So, for example, if a geographical reviewer was somewhat puzzled by the non-linear style of Tim Cresswell's *Maxwell Street*, finding it to be too tangential, then a more literary reviewer compares it to other creative writing:

> the structure of the book is intentionally nonlinear, but instead of the smooth, poetic flow one encounters in other words (like the writings of Maggie Nelson or Rebecca Solnit, both of whom are quoted in this book), its a bumpy read.[107]

Such comments draw attention to aesthetic form, but also raise questions about whether the bumpiness of the non-linear style was doing certain kinds of work, after all, non-linear accounts of place might not seek smoothness? Or, was this just 'bumpy' writing? Elsewhere, Neal Alexander frames these conundrums for us in his review of David Matlesses *Regional Book*.[108] Alexander, a poet and literary scholar who works on conceptualising literary geographies, offers a wonderfully reflective review in which he dramatises on the page the questions he faced as a reviewer.[109] He writes,

> anyone coming to the discipline without a prior knowledge of cultural geography would be amazed by the lack of a 'neutral/objective' voice, lack of argumentative rigour and even of argument and lack of institutional academic apparatus such as references… but what strikes me most forcibly in relation to the Regional Book is how Matless's geographical descriptions resemble prose poems, even if they aren't explicitly presented as such.[110]

Alexander continues this reflective tone, asking,

> what makes me read them as prose poems? Almost everything about them: the deliberately unexpected selection of detours. The parataxis whereby items from disparate registers and discourses are brought together in adjacent sentences, or even in the same sentence… the readerly experience of leaving each piece simultaneously impressed, amused and stimulated on the one hand and curiously underfed on the other, as if the blow glanced off but we called touch anyway.[111]

Alexander is a reviewer with the vocabulary, skills, and consciousness of the prose poem to recognise that form, even if not 'explicitly presented as such.'[112] He draws on his own experience of practice to illuminate what he reads on the page,

> there was a little worry in my mind as I read this book, does this feel a bit like a writing workshop? But actually I think Matless pushes things so far it becomes a legitimate style, not do you need this word? but "what's added by taking this word out?"- a highly significant difference.[113]

Yet he also reflects on his limitations as a reviewer, 'whether this book contributes to an academic discipline, I'm not equipped to say. But as landscape writing, it is smart, rich, and memorable. It also gets me thinking about the prose poem as a vehicle for writing about landscape and history – what is it that makes it attractive?'[114] Encountering Alexander's review is exciting in terms of its explicit framing of questions around how we 'read' those works that sit at the intersection of geography and creative practice. He situates front and centre questions about how we manage and handle the limits of our individual and collective disciplinary knowledge and practice experience. His open reflection on the work and his limitations in the face of it, draws out not only a sense of multiple forms of expertise that these texts demand, but also the need to challenge our engrained assumptions about texts and what they should do. In short, Alexander's review presents, for me, a clear need to evolve new forms of critique for these ways of working. These are forms of critique that, as JAR's editorial reflexivity makes incredibly clear, have front and centre less protocols for review, than an openness to what happens in the encounter with each individual text, and the forms of reading and looking that each form might require from us.

Creative practices and citations

> Digital outputs such as followthethings.com risk being bypassed by more traditional practices of academic review, and our insistence that it should 'stand on its own' without accompanying academic papers doesn't, admittedly, help. Ian Cook et al.[115]

In this final part of the discussion of us as 'readers' and audiences for creative work in geography, I want to reflect on another role we have in relation to the page, as scholars who reference or otherwise cite work. Clearly producing articles and books in relation to practices not only ticks boxes on CVs but also enables the exposure of audiences to work, both directly and in the form of citations. Indeed, journals like JAR clearly have as their driving force a desire to ensure that practice-based research can fully participate in cultures of journal publication. Where creative works lack an academic paper form, their presence within journal articles, and thus their circulation is, as Ian Cook notes above, challenging. As such, thinking through how it is exactly that we might cite work at the intersection of creative practice and geographical research is worth dwelling on further.

Vociferous academic debate surrounds any assumption that citation should be a proxy for impact; relevance; importance; and, of course, quality. Citation, together with the still important journal in which a paper appears, has implications for job prospects, promotion, and other forms of performance evaluation, but also for how it is that some voices are represented and included over others in intellectual conversations. Citational practices have come under increasing scrutiny of late as geography reflects on the politics of its knowledge cultures and as policies such as DORA, the Declaration on Research Assessment, become more widely discussed.[116] For Sara Ahmed citations are a practice through which we choose how to reproduce our discipline.[117] As such, they are a performative technology of power that – as Carrie Mott and Dan Cockayne note – contributes to the 'uneven reproduction of academic and disciplinary geographic knowledge.'[118] While, rightly, much work on the politics of citation foregrounds questions of gender, race, and ethnicity, these discussions also have purchase onto questions of the kinds of work that gets valued. Following Mott and Cockayne, we need to ask not only how can we 'rethink citation as a progressive technology rather than one that serves to make invisible particular bodies and voices'? but also how can we rethink it so it does not make invisible certain kinds of work.[119] As they write at length,

> At the level of our universities and professional institutions, it is important to include a wider range of practices in our definitions of academic

'dissemination' such as teaching, other forms of undergraduate engagement (e.g. field trips, community outreach), conference presentations, public talks, university service, newspaper articles, interviews, recorded talks, online blog posts, and artistic or multimedia projects.[120] While it can be challenging to cite beyond a narrow range of acceptable forums, it is crucial that we do so in the interest of legitimating the multiple ways that knowledge is produced. We should de-emphasize the importance of 'measurable' outputs: citation cannot be adequately used as a proxy for quality, authority, or impact.[121]

The so-called humane system of using endnotes or footnotes rather than in-text citations of course enables the more easy citation of other kinds of work, but such a system is increasingly dying out. Not only do many geography journals require the use of in-text citations, there also seems to be increasing moves to discourage the use of footnotes and endnotes. Digging into the citation reports I receive as editor of *cultural geographies*, it is clear that the 'cultural geographies in practice' pieces display much the same kind of citation profile as our more 'standard' paper. In other words, very variable. Looking more closely though, it is clear that those pieces with more text tend to gain more citations than visually driven pieces or artist's pages. Interestingly though, alongside citation reports we now receive reports on Altmetrics. Altmetrics are akin to citations, but sit at the interface of academic work and social media.[122] Altmetric.com uses a paper's DOI to scan online sources (e.g. blogs, Twitter, Facebook, etc.) for mentions of scholarly papers. For social media mentions (e.g. Facebook, Twitter, and Google+) they look for links to scholarly papers within publicly available posts. For blogs, they search for links to scholarly papers in a list of more than 3,700 blogs individually reviewed and added to the Altmetric.com database by a 'data curator.'[123] Altmetric.com also collects online mentions of scholarly papers from reports published in mainstream news outlets and magazines as well as the RSS feeds of news sites. Interestingly, one cultural geographies paper had no academic citations but had one of the highest Altmetric scores. Whilst those in favour of these forms of metric value their immediacy, their measurement of diverse engagement and relevance. Others, naturally, have significant issues with these same aspects, as well as the so-called 'click-bait' effect that is enhanced in these online tools. Further, they still foreground the need for a DOI – thus not yet making space for those forms that would not have a DOI. As such then, those of us working at the interface of geographical and creative practices need to think further about how it is that our published work can take account of the breadth of practice and scholarship that influences us. How is it that we might resist and even counter-act the skewing of academic publications towards other academic publications in their discussion of influences and contributions? For space needs to be made to ensure that when doing our 'reviews' these might be both literature and practice, and when drawing out inspirations and where our work might be influential we can take account of other publications as well as creative forms.

Creative pages

The relationship between geographical research and creative practices is one that produces, as the preceding chapters have hopefully made clear, a plethora of different forms of output, whether this be artefacts, including sculptures, paintings, or whether this be a series of events, from performances to dance work, as well as installations. Yet, amidst all of these forms, the page as a site for the production of geographical knowledge persists. Yet it persists in ways that suggest that the form of the page; the kinds of contexts (books, online fora, alternative publication contexts) it might be found within; as well as how it is we engage it as producers, as editors, as reviewers, and as readers, might be being reshaped. As a range of historical geographers studying geography's books and diverse print cultures have made clear, the printed form geography takes on the page has a productive, co-constitutive relationship with/to the kinds of geographical knowledge that is made. As such it is interesting to trace some of the forms that the pages of geographical knowledge now take. This chapter has aimed to do so through encounters with a series of pages, as producer, as reader, as reviewer, and as reader of reviews. In doing so, it becomes clear that the print-spaces and online-page-scapes emerging at the interface of geography and creative practices are forcing us to ask a series of questions about the page. This includes, what we want on it, how it might be considered a site of negotiation between the demands of creative practices and of geographical audiences, and about how we are experimenting with what might be possible to do on the pages of our geographical texts. Yet of course, we need, as historical geographers made clear, to look up from the detail of the pages and the material forms of a book-work and its aesthetics, to reflect on the wider geographies of circulation that these pages are embedded in. We see how sections of established journals are important sites for the visibility and reputation of creative practices and practitioners. Further in an era when there are concerns over the demise of the monograph, other forms of geographic book work are perhaps growing, whether this be artists' books, zines, or work with small presses. Sites of review – whether guidelines to peer-review or published reviews – offer interesting points of reflection on the cultures of reading that these creative pages require and how the discipline might need to reflect further on this. Whilst the politics of citation becomes crucial in these discussions, where the role of citations as a primary technology of academic knowledge production is a technology that seems less than well suited to ensuring the circulation of these ways of working. The implications this has for the reach of these creative pages, and the reputation of those producing them should tend the minds of those of us who not only produce, but also edit and review these kinds of work. Of course, if the page is one of the normative sites for the consumption and circulation of geographical knowledge, then for many forms of creative practice the exhibition has been equally important. It is to the form of the 'research exhibition' that discussion now turns.

Notes

1 Mayhew, "Materialist Hermeneutics."
2 Ibid.
3 Blackmore and Kandiko-Howson, "Motivation."
4 Mayhew, "Materialist Hermeneutics"; Ogborn and Withers, *Geographies of the Book*.
5 Keighren, *Bringing Geography to Book*; Keighren, "Geographies of the Book"; Keighren et al., *Travels into Print*.
6 Keighren, "Geographies of the Book"; Ogborn and Withers, *Geographies of the Book*.
7 Livinsgton, "Science, Text and Space," p. 13.
8 Lippard, *Six Years*; Drucker, *The Century of Artists' Books*.
9 Phenlan, *Unmarked*.
10 Lippard, *Six Years*; Drucker, *The Century of Artists' Books*.
11 Meinig, "Geography as an Art."
12 Olsson, *Birds in Eggs/Eggs in Birds*; Pred, *Recognizing European Modernities*; Cresswell, *Maxwell Street*; De Leeuw, *Geographies of a Lover*; De Leeuw, *Skeena;* Magrane and Cokinos, *The Sonoran Desert*.
13 Ward, "The Art of Writing Place."
14 Alfrey et al., *Art of the Garden*; Hawkins and Lovejoy, *Insites*; Tolia-Kelly, "SPILL"; Cresswell *Soil*; Cutler, *Nostalgia Forest*; De Leeuw, *Geographies of a Lover*; De Leeuw, *Skeena*; Hoskins and Whitehead, *Mining Memories*.
15 Magrane et al., *Geopoetics in Practice*.
16 Crang and Mitchell, "Editorial."
17 https://archives.history.ac.uk/makinghistory/resources/articles/HWJ.html (accessed 12/2/2020).
18 https://www.tandfonline.com/action/authorSubmission?show=instructions&journalCode=rgeo20 (accessed 12/2/2020).
19 https://www.elsevier.com/journals/emotion-space-and-society/1755-4586/guide-for-authors (accessed 12/2/2020).
20 Data assembled through analysis of the *Cultural Geographies* online archive of all articles. https://journals.sagepub.com/loi/CGJ (accessed 12/2/2020).
21 DeLyser and Hawkins, "Introduction"; Volume 21, issue 1 of *Cultural Geographies*; Sachs-Olsen, "Performing Urban Archives," Volume 23, issue 3 of *Cultural Geographies*.
22 Phillips, "Fieldwork."
23 Lovejoy, "…here nor there…"
24 Mayhew, "Materialist Hermenutics."
25 Ibid.
26 Ibid.
27 Ibid.
28 Ibid.
29 Cook et al., "Afters," p. 135.
30 Sachs-Olsen and Hawkins, "Archiving an Urban Exploration."
31 http://www.perditaphillips.com (accessed 12/2/2020).
32 Ibid.
33 Phillips, "Doing Art and Doing Cultural Geography."
34 Ibid.
35 Mayhew, "Material Hermenutics."
36 See for example, https://blogs.lse.ac.uk/impactofsocialsciences/2019/08/05/do-monographs-have-a-future-publishers-funders-and-research-evaluators-must-decide/ (accessed 12/2/2020); see for example, "Researcher's perspectives on the purpose and value of the monograph: Survey Results 2019, produced by Oxford University Press and Cambridge University Press; https://global.oup.com/academic/pdf/perspectives-on-the-value-and-purpose-of-the-monograph (accessed 12/2/2020).

37 https://landscapesurgery.wordpress.com/2015/02/03/measuring-the-value-of-monographs/ (accessed 12/2/2020).

38 Lovejoy and Hawkins, *Insites*.

39 Pred, *Recognizing European Modernities*; Olsson, *Birds in Eggs/Eggs in Birds*.

40 Cresswell, *Maxwell Street*; Krupar, *Hot Spotter's Report*.

41 Cutler, *Nostalgia Forest*; De Leeuw, *Skeena*; De Leeuw, *Geographies of a Lover*; Cresswell, *Soil*; Kichin, "Engaging Publics."

42 Thrift et al., *Patterned Ground*; Pile, *City A-Z*.

43 Magrane and Cokinos, *The Sonoran Desert*.

44 See for example, Cutler, *Nostalgia Forest*; De Leeuw, *Skeena*; De Leeuw, *Geographies of a Lover*; Lovejoy and Hawkins, *Insites*; Parrott, *Swallet*, details at http://www.floraparrott.com [last accessed 22nd June 2020].

45 Lippard, *Six Years*.

46 Drucker, *The Century of Artist's Books*.

47 Lippard, *Six Years*, p. 45.

48 Drucker, *The Century of Artist's Books*.

49 Ibid.

50 Ibid.

51 Ibid.

52 Ibid.

53 Drucker, *The Century of Artist's Books*; Lippard, *Six Years*, p. 50.

54 Drucker, *The Century of Artist's Books*.

55 http://www.colinsackett.co.uk/uniformbooks.php (accessed 12/2/2020).

56 Ibid.

57 DeSilvey et al., *Anticipatory History*.

58 Bond et al., *Visible Mending*; Matless, *Regional Book*; Neate and Craggs, *Modern Futures*.

59 Newman, "Authorising Geographical Knowledge."

60 diffusion.org.uk (accessed 12/2/2020).

61 Ibid.

62 diffusion.org.uk; Lorimer and Foster, "Cross-Bills"; Yusoff, "Landscape."

63 Pratt, *Its Space Jim*, http://diffusion.org.uk/?p=97 (accessed 12/2/2020).

64 Duncombe, *Notes from the Underground*.

65 Ibid.

66 Ibid.

67 See Bagelman and Bagelman, "Zines"; http://www.camps2cities.com/project/zine/ (accessed 12/2/2020); https://everydayausterity.wordpress.com/exhibition/ (accessed 12/2/2020).

68 Bagelman and Bagelman, "Zines," p. 2.

69 Ibid.

70 Bagelman et al., "Cookbooks."

71 http://www.camps2cities.com/project/zine/ (accessed 12/2/2020).

72 Ibid.

73 Ibid.

74 https://everydayausterity.wordpress.com/exhibition/ (accessed 12/2/2020).

75 Ibid.

76 Cresswell, in Merriman et al., "Landscape, Mobility, Practice."

77 Bishop, *Installation Art*; Hawkins, "The Argument of the Eye."

78 Bishop, *Artificial Hells*; Harvey, *Fair Play*.

79 Keighren, "Geographies of Reading," p. 20.

80 Livingstone and Withers, *Geographies of Nineteenth-Century Science*, p. 9.

81 Barnett, "Disseminating Africa."

82 See summary in Keighren, "Geographies of Reading."

83 https://www.jar-online.net (accessed 12/2/2020).

84 https://www.jar-online.net (accessed 12/2/2020).

85 https://jar-online.net/journal-artistic-research (accessed 12/2/2020).
86 https://www.jar-online.net (accessed 12/2/2020).
87 https://jar-online.net/journal-artistic-research (accessed 12/2/2020).
88 https://www.jar-online.net (accessed 12/2/2020).
89 Holert, "Evaluating Knowledge= Evaluating Art?"
90 Ibid.
91 Baxter, "Review of Modern Futures."
92 Daniels, "Review of Bipolar," p. 425.
93 Keighren, "Geographies of Reading."
94 Livingstone, "Science, Text and Space."
95 Rupke, "A Geography of Enlightenment."
96 Daniels, "Review of Bipolar," p. 424
97 Hubbard, "Modern Futures," p. 1; Baxter, "Review of Modern Futures," p. 654.
98 DeSilvey, "Mining Memories," p. 387.
99 DeSilvey, "Mining Memories," p. 378.
100 Hubbard, "Modern Futures."
101 Edensor, "Review: Visible Mending."
102 Pickles, "Radical Thought-in-Action."
103 Linder, "Maxwell Street."
104 Worpole, "Nowhere to Run, Nowhere to Hide."
105 Beer, *Darwin's Plots.*
106 Livingstone, 'Science, Text and Space," p. 393.
107 Alexander, N (2015) Seeing More Flatly. Cultures of Place Blog. https://cultures
    ofplace.weebly.com/blog/seeing-more-flatly (accessed 12/2/2020).
108 Alexander "Seeing More Flatly."
109 Ibid.
110 Ibid.
111 Ibid.
112 Ibid.
113 Ibid.
114 Ibid.
115 Cook, "A New Vocabulary."
116 Mott and Cockayne, "Citation Matters," https://sfdora.org (accessed 12/2/2020).
117 Ahmed, https://feministkilljoys.com/2013/09/11/making-feminist-points/ (ac-
    cessed 12/2/2020).
118 Mott and Cockayne, "Citations Matters."
119 Ibid., p. 966.
120 Ibid.
121 Ibid., p. 968.
122 Hammarfelt, 'Using Altmetrics."
123 Ibid.

8

EXHIBITION

Four research exhibitions

If since the mid-twentieth century the exhibition has been '*the* medium through which most art becomes known,' this chapter explores the growth of exhibition making at the intersection of geography, art, and research.[1] Geography is rarely included amongst the key group of modern 'exhibitionary' disciplines. In other words, those disciplines, such as anthropology, geology, and art history, whose evolution was closely bound up with the evolution of sites of display.[2] Yet geographers have recently been reflecting on the discipline's rich exhibitionary histories, and exploring the important role exhibitions have played in the production and circulation of geographical knowledge.[3] The Royal Geographical Society in London has been a key site for these historical explorations, with studies demonstrating the vital role of visual lectures, lantern slide shows, and a photography gallery in the evolution of the Society and the forms of geographical knowledge it sought to foster and promote.[4] Furthermore, as well as reclaiming geography's exhibitionary histories and the role of the curation of objects and knowledge in the founding of the modern discipline, we must also consider the growth of exhibition-making as a contemporary geographical practice. Indeed, we could perhaps claim geography as an emerging exhibitionary discipline – the focus of which is the geographical research exhibition.

The term 'research-exhibition' has become increasingly common in recent years, both through the growth of practice-based research – resulting in works – but also in the growth of exhibition-making as a practice of research communication.[5] For many within the art world the emergence of the research exhibition is an extension of the twentieth-century evolution of exhibitions as forms of critical enquiry and knowledge production in their own right, rather than merely containers for artworks and objects.[6] An often cited origin for the phrase

FIGURE 8.1   *Blood Bricks,* The Building Centre, London, 2018. (Image, Royal Holloway University of London)

FIGURE 8.2   *Forensis,* MACBA, Barcelona, 2016. (Image, author's own)

'research exhibition' is a 2003 conference paper by design scholars Chris Rust
and Alec Robertson in which they detail the process behind creating an exhibition as a research publication.[7] They were driven by a desire to probe how the
exhibition might offer a viable form of research dissemination in cases where
artefacts are produced and thus where communication through writing felt limited. Craft research scholar Kristina Niedderer and colleagues evolved Rust and

FIGURE 8.3 *Basecamp*, London, 2017. (Image, author's own)

FIGURE 8.4 *The Artificial Cave*, David Roberts Foundation, London, 2016. (Image, author's own)

Robertson's paper into a more programmatic model, again placing emphasis on the exhibition as a form of research dissemination.[8] Yet the exhibition-as-research-communication model has since been evolved to enable more conceptual thinking about what a research exhibition could be. A 2010 symposium at Tate Modern, London 'Beyond the Academy: Research as Exhibition' for example, jointly organised by the Tate Research Department and LCASE, a UK Arts and Humanities Research Council Funded doctoral training partnership between a number of London universities, queried the terms of the relationship between research and exhibitions.[9] It assembled a panel of curators, academics, and artists, many of whom occupied more than one of these roles, including geographer Felix Driver, who explored both what research might bring to practices of curation and whether or not the idea of the exhibition was being 'distorted or creatively extended by new disciplinary practices and knowledge.' In what ways, the panel explored, 'do new forms of research exhibitions create new types of knowledge and experience for the audience?' It is against this backdrop of the open questions that this chapter's focus on the geographical research exhibition takes place.

The growth of a geographical research exhibition is perhaps not surprising given the growth in the interactions between geographical research and both cultural institutions and creative disciplines. Geographers have long been working in close proximity to cultural institutions large and small, often with a significant gallery component. There is now a series of national and international organisations whose research and programming bears the mark of geographers' research topics and practices, such as Kew Gardens (UK) The Neon Gallery (USA); the domes of Biosphere2 or the landscapes of the Desert Museum in Arizona; Cape Farewell (who take artists and scientists on boats to the Arctic); as well as the Bundanon Trust, Australia.[10] Further, much of the artwork produced at the intersection of geographical research and art is presented in exhibitions, whether group or solo shows. Other examples of this work is held in the permanent collections of major institutions around the world, while further examples go on tour or features in some of the world's major arts and architecture festivals as well as being installed in a range of temporary spaces, from hipster pop-ups to provisional institutional spaces in departments and university buildings. In some cases the art works shown, and the exhibitions produced are invited and funded by the institution, in other cases geographers hire venues and pay for exhibition designers to help showcase their research. Further, and as the earlier chapter on 'thesis' explored, many doctoral candidates present an exhibition as part of their viva. We might claim too, the emergence of the figure of curator-geographer. We witness for example, a number of curators of major institutions holding Geography PhDs, including those working at the Science Museum, the British Library and the Tate in London.[11] We also find Geographers acting as curators of public programmes and major architecture biennales as part of their research, and taking up ongoing roles in helping to shape the direction of arts and cultural organisations.[12] Geographers also work as co-curators, in collaboration with a series of institutions from major global museums, to smaller local or regional institutions/organisations to produce exhibitions and events.[13] Geographers are also working as curators of exhibitions made of participatory work, or about participatory practices, of the sort discussed in Chapter 4 on community.[14] We also find, interestingly, a growth in departmental exhibitions and university exhibition spaces, offering the chance for in-house curation as well as on-line exhibitions linked with research projects and journals, and as a longer-term record of more ephemeral installations.[15] There are also geographers like Karen Till, based at the University of Maynooth, Ireland, whose geographical research is thoroughly enmeshed with a curatorial practice. Key amongst her diverse curatorial practice is her role in producing exhibitions associated with the networks of artists and geographers she has helped to create, including *Mapping Spectral Traces*.[16] Consistently appearing in geography departments and at a range of conferences and other institutional spaces Karen's practices have been an important part of the increased visibility of creative practices within geographical research.[17] This rich ecology of geographical-curatorial practices that has emerged over the last few years is both a product of and has helped produce the expanded field of curation such that while the skill of the curator is still valued and exists as a specialist role, many artists will now curate their own and others' work.

Geography has had and continues to have a rich exhibitionary landscape. Against the backdrop of still nascent discussions of what is meant by the research exhibition, this chapter queries the emerging form of the geographical research exhibition. It does so through re-staging a series of encounters with research exhibitions, both those I was involved in the evolution of, as well as those I went to see. Importantly, there is a huge range of exhibitions discussed here. Whilst risking a somewhat fragmented approach, such a range offers the chance to reflect on the different tensions and possibilities that emerge from the discussion of a series of exhibitionary forms. One of these, *Blood Bricks* (2018) (Figure 8.1), was an exhibition of photojournalism commissioned as part of a geography project led by social and development geographers where images were given a particular epistemological force, understood to visually reveal 'untold stories' and to help to enhance research impact. Another point of focus are the arts exhibitions produced by the research organisation, Forensic Architecture (Figure 8.2). These exhibitions, situated in international gallery spaces frame a negotiation of tensions amidst art, research, and politics, and raise questions of what and how our research should be put on display. Finally, there are a series of smaller exhibitions produced as part of practice-based research (Figures 8.3 and 8.4). Here questions are raised around what exactly, if anything, is being put 'on display' in these contexts, exploring instead how a research exhibition might offer a form of 'research-in-action.' The chapter is built through first-hand accounts of exhibition experiences, as both an audience member and, in some cases, as part of the team working on exhibitions, which are combined with material from interviews and secondary texts. What becomes clear is that research exhibitions are a hugely diverse form, encompassing many different relations between geographical research and exhibitionary practice. In the context of this evolving geographical exhibition landscape with its increasingly complex ecologies, the exhibitions that do exist raise a series of powerful and important questions about putting research on display. These include, how and why geographical 'data' is being put on display, how is research made present in exhibition spaces, including through art works produced through practice-based research, what aspects of the research are put on display, and what is left out? How is data transformed over the course of a research exhibition, and what does that mean for research and for how exhibitions are produced within it. As with the other chapters, this is not about foregrounding a singular sense of the research exhibition, but rather using grounded examples and experiences to explore these pertinent questions.

Exhibiting research 1: putting data on display

Exhibition A

'Blood Bricks: The Untold Story of Modern Slavery in Cambodia.' The exhibition takes place in London's Building Centre, an architectural resource centre-cum-cafe-cum commercial space-cum exhibition venue. The main route into the exhibition takes you through the trade booths of firms selling insulation, paint, bathroom fittings, bricks, and other masonry. The focus of the exhibition was a series of photographs, and their visual telling of the story of debt-bonded modern slavery associated with Cambodia's brick kilns, and the role of climate change in exacerbating these practices and deepening the plight of many families. The images filled the relatively small space, the rich colours and textures of the kiln fires, of moody storm clouds over fields scarred by soil erosion created a very atmospheric space. Grouped according to three spatial themes: City, Kiln, Field, the short captions accompanying the images detailed the research findings through these sites.

Image and research seem thoroughly entangled from my entry into the exhibition, where an infographic details the findings in pictogram form. I wondered at the choice to display such different visual languages together, yet together images, captions, and infographic generate a certain kind of informational aesthetics, harnessing together the visceral power of the images and the detail of the stories behind them. The exhibition was also the launch event for a longer policy report on the research, a highly professional brochure that used some of the photographs to create a striking visual document that was available for consultation in the exhibition space.

Field notes from exhibition visits, Autumn 2018

Exhibition B

The striking opening image, filling the screen of my laptop, is a silhouette of a young girl's back, her long ponytail, and her gloved hands. Her fingers laced into holes in bricks as she walks into the light, a cart of bricks in front of her, piles of sand in the background. 'Modern-Day Slavery in focus,' a white font on red, and then in larger print on black, 'How climate change forces farmers into slavery in Cambodia's brick kilns - in pictures.' The presentation is striking, the black background, the scale of the image. A small caption off to the

FIGURE 8.5 Modern–Day slavery in focus, in pictures.[18] (Image, Royal Holloway University of London)

left said, 'A 10-year old girl helps her parents by carrying fired bricks out of the kiln Photograph: Thomas Cristofoletti/Ruom/2018 Royal Holloway/ University of London.' Scrolling down the page, I see the date and time of the story's posting, Thursday 18th October, 2018, and the authors' names – Hannah Summers and Thomas Cristofletti. Summers's name in a different colour, you can click through to her other articles, including travel pieces and work on the state of the oceans, and issues of harassment of female barristers. Continuing to scroll down the page, there are twenty-six images in all, including a repeat of the title image. Each filling the screen aside from a few inches of black either side and a short caption. The images are large, you can appreciate the colours, especially against the black, you are enticed to further looking and the images reward this. At the bottom of the screen, are those donors who support The Guardian online image galleries, in this case, Humanity United. The latter is a philanthropic organisation backed by E-Bay Billionaire Pierre Omidyar, known for his work in for-profit and non-for-profit online journalism (Figure 8.5).[19]

Exhibition C

Entering the website for the online blog of the journal Environment and Planning D: Society and Space, the title – in block capitals on a white page – reads, 'Blood Bricks: Untold stories of modern slavery and climate change from Cambodia, By Katherine Brickell, Laurie Parsons, Nithya Natarajan and Sopheak Chann. Photographs by Thomas Cristofoletti.'[20] Scrolling down, the text begins by introducing the exhibition, it starts – 'In October 2018, the research-led exhibition,' it ends by saying, 'this photo essay provides a more permanent home for the exhibition.' The text goes onto detail the context for the exhibition and the launch of the research findings report, the widespread media coverage the exhibition and report received, and the wider research context. Scrolling down the page you find the first photograph, restricted in size by the blog format the photographs seem lost on the white page. Their aesthetic force far less compelling for this viewer than in either of the other contexts.

Blood Bricks – three exhibitionary contexts

In 2018, I worked with the Blood Bricks team to put together a workshop that accompanied their exhibition at the Building Centre.[21] The workshop was funded through the project's original grant from the UK Department for International Development and the Economic and Social Research Council.[22] The workshop was also a response to the launch, in 2015, of the Global Challenges Research Fund, a £1.5 billion fund from the UK Government to 'support cutting-edge research that addresses the challenges faced by developing countries.' GCRF as it has become known often ties its funding to the evolution of research through interdisciplinary approaches, and we were interested to explore the intersection of arts and humanities scholars, and practitioners and social scientists in these grant making contexts. The workshop also offered a chance for me to spend some more time exploring the epistemological questions raised at the interface of social science's approach to addressing challenge-led research and the sometimes conflicting epistemologies of creative practices.

As part of the preparation for the workshop I went looking for examples of other exhibitions that could be considered to be putting 'data' on display. This would include where the photographs, drawings, or videos that constitute the exhibition are ones that have been produced either by professionals such as Thomas Cristofoletti, or by participants, as was discussed in Chapter 4. Interestingly, oftentimes, academics mounting these exhibitions speak of them in terms of enhancing the reach and impact of the research findings, conveying its messages to new audiences or, sometimes, effecting the identities of those communities whose works were being shown. Sarah Hall's exhibition 'Everyday Austerity,' aimed to 'lift the lid on austerity' through the turning of 'first-hand, personal accounts of everyday life in austerity' into drawings by North-West zine artist Stef Bradley which exhibited alongside field notes, sound-bites, and collected materials. The exhibition toured a series of venues in Northern England, receiving wide-spread acclaim including several University prizes for research impact and social responsibility. As the twitter feed for the exhibition records, everywhere from museum's and local heritage trusts, to research centres committed to social activism, welcomed and valued the exhibition and its zine.[23]

As the Blood Bricks team made clear in the online essay, the images were included within the project as

> a way to engage stakeholders and publics in the work, to illustrate the reports and produce an exhibition for the UK and Cambodia to engage discussion about how the Cambodian brick production industry combined with climate change was reducing thousands of citizens to debt bondage.
>
> – *Modern Slavery*[24]

The success of the images in doing this work was indicated through the project's list of the different media outlets the images had reached, and by drawing a direct link between the power of the images and the engagement of in-country policy-making officials with their project findings. The photographs and their exhibitionary potential were clearly a highly effective mediating tool, offering another dimension to the potent stories the team had to tell from their research. Yet, across the three different exhibition forms, three different types of research gaze emerged.

As contemporary exhibition makers it is important that we recognise that expanded nature of exhibition practice. In other words, not only have exhibitions moved beyond the museum's walls, but it is well recognised that the primary site for consumption of visual materials, including artwork, is often online spaces. These might be blogs or specialist sites, or increasingly in the case of art, through social media applications especially visually dominated ones like Instagram. As such, thinking about the exhibition of research is also to appreciate myriad on- and off-line sites variously designed to support the variety of things we might want to display and both the challenges and aesthetic and contextual opportunities these different objects of display pose. Evolving their concept of a research exhibition in design, Rust and Robertson draw on the work of museologist Peter Vergo to explore the place of 'research' in display. Vergo draws a distinction between 'aesthetic exhibitions' and 'contextual exhibitions.'[25] In the former the artefacts (whether historical objects or art-/craft works) are presented with little additional information, the process of encountering them being largely, they explain, experiential. Whilst, in the latter, the interpretation of the displays is framed by 'informative, comparative and explicatory' material. Rust and Robertson are very clear, research exhibitions must be contextual exhibitions, otherwise they will, they claim, be unable to carry out their role to mediate access to research findings. Interestingly, looking across the three exhibitionary contexts for the Blood Brick's photographs, it seems that, in fact, aesthetics and context worked together to create a powerful display.

Each of the three exhibitionary contexts draws on the same subset of around twenty images selected from the sequence of many thousand taken by multimedia journalist Thomas Cristofoletti. Cristofoletti, based in Cambodia, was commissioned by the team to make images alongside the site visits and interviews conducted in the field by other team members. The photographs emerge not just from the team's research context but also from Thomas's history as an image maker. Cristofoletti is the founder of Ruom, a multi-media collective dedicated to reporting social issues across Asia. Similar to the philosophy of Magnum photography, the agency uses multi-media methods – photography, video, digital online content – to bring together journalism, art, and storytelling.[26] Looking at the work they do it is clear that social commentary aimed at driving social change informs the stories they create. This includes the Cambodian Divers who are 'De-mining the Depths' (2013) of the country's lakes and rivers, which are littered with unexploded munitions with little training, or that of 'Blood Sugar' (2013), which, similar to Blood Bricks, explored the lives of workers in Cambodia's sugar fields and the land they lost to the agro-industry. Their aims and practices, including those of working with NGO's, clearly resonate with the interests and ethos of the Blood Bricks team, and their desire to tell these untold stories of modern slavery in Cambodia.

It was this sense of the image as revelatory that was core to all three of the exhibitionary contexts in which I encountered the photographs. This epistemological context has been central to photography almost since the medium's creation – the image as truth-teller. How this was done though, and the role of the photographs in each case was quite different. The Guardian's online photo-gallery, designed to present visual material, offered a lush visual frame for the photographs. A vertically organised blog-like posting enabled images to almost fill the screen with smaller captions off-set to the left as the viewer scrolled down the page. The photos here were presented, perhaps a little disconcertingly, as 'the reality behind the research,' as if the research itself was another story, behind which was a 'reality' these images could get at.[27] The captions are a couple of sentences of often emotive stories of hardship, quotes from participants, from the British Ambassador to Cambodia and the occasional detail about the research project, 'the team made recommendations to the Cambodian Government.' There is a stark contrast between this online presentation and the 'visual essay,' produced for the journal Society and Space's blog. The text and images in this essay are the same as those in the exhibition, front-ended by an introduction, yet the visual presentation is severely restricted by the blog's format. The photographs, captioned here as 'figures,' are small and seem to lose power against the dominating text, which is in a large font; even the captions beneath the images themselves are large. The white background of the post does not help the viewing experience. Set amidst the blocks of words the photographs function more as illustrations, they are present but are hardly eye-catching. Here though, the research context is asserted through text, the opening line informing us that this is a 'research-led' exhibition, while the text goes onto detail the exhibition's context and the launch of the research findings report, and the widespread media coverage the exhibition and report received. In the midst of all this detail the photographs themselves recede. This is in contrast to their presentation in the exhibition of the Building Centre. Here in the small low-ceiled space, not exactly the white cube gallery, the images shine out from the walls, their colours are intensified, their large scale is engaging, the landscapes are detailed, by turns vibrant and moody. The experience is engaging. The use of hanging panels and large images which fill walls offers a more immersive experience, thrusting the images into the space in a more striking way than had they been in frames on walls.

Across the three exhibitions the images take very different roles. Sometimes directing the audience experience, other times as more illustrative of the points made by the text. Each context asserts the 'researchness' of the photographs in a different way. This is less a detailing of their methodological production in any context, but rather uses a range of strategies. In the Building Centre exhibition it is the infographic as much as the text panels that creates a sense of the 'research' context from which the photographs emerged, as an integral part. Whilst in the blog essay, the text dominates, the photographs discussed but less viewed. Here the photographs, situated as part of an impact strategy, seem less the research than its dissemination tool. Whilst, in the Guardian essay, it is really the association with Royal Holloway, University of London, and the occasional mention of the project that serves to frame the images as research, rather than anything else. Framed as the 'reality behind the research project' the photographs seem more distanced from the research than in the other contexts.

What emerges from comparing these exhibitionary contexts is a clear sense of exhibition as a tool for mediation, but also a sense that in putting data on display, attention needs to be paid to the negotiation of the status of the materials on display as 'research.' Clearly different contexts require the images to do different forms of work for a range of audiences, so the relationship of image and research is multiple, rather than singular and fixed. What is needed for a journalism context is different to a more academic essay and different again from a commercial gallery space and an exhibition designed to appeal to policy-makers. In some cases the 'research' context of the images is a point of value, in other cases (the Guardian) it is far less important than the wider stories the images tell.

Exhibiting research 2: creating a research aesthetic?

Walking into the first gallery, a huge collage curves across the wall in front of me 'Forensic Architecture: Towards an Investigative Aesthetics,' the blocky text superimposed over it reads. The gallery walls are covered with materials, a wall of organised photographs, highlighted maps, coloured satellite images and aerial photographs. 'Thresholds of Detectability,' a large protruding wall announces, the sort of spatial equivalent of a text box in a text book. 'Forensis' another reads, a long paragraph of text, in the middle of which, highlighted text summarises 'Forensic Architecture seeks to unlock the potential of forensics as a counter-political practice. Inverting the direction of the forensic gaze.' The gallery leaflet in my hand observes, 'Forensic Architecture has not only shed new light on human rights violations and state crimes across the globe, it has also given rise to a new form of investigative practice, to which it has given its name.'

It feels a bit like being in a visual essay, there is a lot of text. It is less a case of looking at an image and then reading a caption and rather more of reading through the text then looking at the image. The normal provision of gallery benches in the middle of the room is rendered somewhat useless for the kinds of encounter that are needed with the volume of information that covers the gallery walls – and anyway if you sit on them you can neither read the text nor really see the films. Not far away a fake swan back chair has been placed seemingly at random. While I am there one man picks it up moving it closer to a panel of text, he carries it to the next room with him. This much information needs a place to pause, to watch, look and read.

In a second gallery there are vitrines – part map-table, part object-case. In some cases the aerial maps remind me of the kinds of war scenes pictured in films, with generals using long wooden sticks to move model troops across the field of battle. Here maps of towns are time-sequenced, logged against the changing forms of clouds, composites of images from satellites, smartphones, drones, and aerial kites, enabling the recreation of precise movements of people and planes on the day of the destruction in question. This is an exhibition that requires you get close in. As the exhibition unfolds the organisational forms become clear, a series of project-based geographically located case studies of armed conflict and environmental destruction, each with text boxes, collages of text, images and often maps, sometimes a video or a sound piece- several times large installation works with multi-media elements.

I spend a long time moving between the spaces, moving back and forward from text to image, object and video. I find myself sitting in front of videos and listening through headphones for many minutes. I develop a pattern of engagement. I approach, look at any images, objects, or installations first, then I move in to read the details, to appreciate the context given for the case studies. Sometimes the explanations are long, not only detailing how the investigations proceeded, but also the effects they had.

Constructed from field notes of exhibition visits
Museum of Contemporary Art, Barcelona, June 2017

Art Historian Claire Bishop asks us to consider how artistic research is presented by audiences, how they experience what she terms the 'informational aesthetic' of this work. One of the trajectories she explores for such audience experiences begins from Lewis Hine's prolific early twentieth-century photography of underprivileged groups in the US. She observes, 'long descriptive captions, which included measurements and statistics which he would have the immigrants, workers and child labourers sign.'[28] By contrast, she also muses on more recent forms of informational aesthetics premised on the installation of volumes of material. Such experiences risk, Bishop reflects, seeming more of an attempt to prove a research process has occurred, such that the audience experiences information overload amidst 'drifts' of material. This is an experience, she suggests, akin to that of internet browsing, 'passing over … skimming, darting here and there, never necessarily alighting on any one thing for too long'; this is, she suggests, research 'as search.'[29] In other cases, however, the 'drift' of material on display might enable the audience to appreciate multiple perspectives in place of the singular master narrative a gallery or museum might normally present.[30]

If my experience of *Forensis* was one of consuming a lot of data, this was in contrast to Bishop's worries over data-drifts and information overload. Instead my experience was more akin to being installed within a visual essay, where the relationship between text and work in front of me was integral, but rather than this being consumed on a page, I was immersed within a kind of research aesthetic.

Text, both its volume and content, was a consistent site of concern in my discussions with the producers of research exhibitions. Akin to the discussion of textual imperialism in both of the previous chapters on thesis and page, there was a clear concern on the part of those evolving research exhibitions to balance the need to minimise the text present but also to prove the exhibition's research credentials, something the text was seen as vital for doing. Geographer Alison Williams reflects on the dilemma of how much of her geographical research to use in the exhibition of artwork produced during the Leverhulme AIR residency collaboration with Matthew Flintham (discussed in Chapter 5). Of the exhibition *Martial Heavens* they produced and then installed in Newcastle University's gallery spaces she reflects,

> we did not want to make it academically heavy, we had a lot of conversations about whether we would have a guide or a booklet, or the idea of doing a newspaper front-cover and having a sort of hand-out at the exhibition. So there was lot of conversation about how we brought the academic stuff into the exhibition... We wanted to have something people could take away with them, but we ranged quite a lot between something that was quite chunky and extensive and relied upon a lot of my research, and what we ended up with which is sort of 3/4 page of text a couple of links to our website, and the rest of it its quite straightforward explanation, so again this was about making it as accessible to people as possible. We wanted people to engage with the art works without having to deal with complex academic ideas, if I start bunging in performativity people are going to you know glaze over ... part of the whole premise of the project was to make these intangible, invisible places accessible so people could see them and understand them, and not obscure them with lots of complex theoretical language about the invisible, their performance, their performativity.[31]

A recent guide for researchers interested in curation produced by the UK Royal Geographical Society and the Arts and Humanities Research Council suggests,

> it is worth bearing in mind that visitors tend to read relatively little in exhibitions and that their reading age is reduced in the gallery environment. More than 80 words on a text panel, for example, and the text is likely to remain unread.[32]

Elsewhere, debates about gallery text have noted that the once discreet placement of text on gallery wall labels focussed on salient details – work, title, date, artist, location – has been replaced by a proliferation of written materials whether downloadable content accessed through QR codes, or apps on smartphones, tablet guides augment the long-present audio-guide with written text, and there is often the ability to annotate and mark up one's gallery experience as you go. This is of course not to mention the now almost ubiquitous social media commentary – restricted to the character counts of Twitter and the Instagram caption or the empheral form of the story or tik-tok video. Whatever the form of text involved, evidence suggests that audiences often spend as much time in galleries considering the text as they do they on the artworks; and further, that the text can come to 'author' the way we look.[33] As such, the text not only imparts important information about the works we see, but shapes how we look,

conditions the pace at which we move through gallery spaces, and can even enable us to escape 'the difficult task of looking' or otherwise engaging.[34]

In the case of *Forensis*, there was clearly much more than eighty words of text.
Indeed, it felt like the text was an exhibit, rather than supplementing the visual
material. Further, it seemed integral to authoring the means of looking, what
for me at least felt a bit like the shaping of a research gaze. Forensic Architecture (hereafter FA) are a research group of architects, visual artists, documentary
film-makers, and multi-media practitioners based at the Centre for Research
Architecture at Goldsmiths University, London. Much of their work has been
funded by large research bodies such as the European Research Council.[35] FA
offers a high profile and provocative example of the powerful coming together of
creative practice and research. Their exhibitionary practice has become a sometimes controversial flash point for debating their work, including *Forensis*, at the
Museum of Contemporary Art Barcelona (2017); their exhibition at the Institute
of Contemporary Art in London (2018); and the inclusion of their work in the
Whitney Biennale in 2019.[36] The exhibition at the Museum of Contemporary
Art Barcelona was one of their first major exhibitions, and was quickly followed
by a nomination for the 2018 Turner Prize, one of the UK's most prestigious
contemporary art prizes.[37] In response to their nomination for the prize, the
group issued a statement saying they were honoured and surprised and did not
think of themselves as artists.[38] This was an interesting statement, given their
choices to display their work in major art galleries, it also brought further critical
attention to concerns that the group had ended up aestheticizing violence, producing an exploitative form of disaster aesthetics.[39] As one review put it,

> the use of critical evidence for eye-catching exhibits for the entertainment
> of gallery-going art consumers is questionable, and becomes more trou-
> bling as FA gain prominence in the art world… their investigative prac-
> tice is at risk of becoming commercialised as a particularly grisly disaster
> aesthetic.[40]

In the context of such concerns, and the ethical importance of their research. I
am concerned to query how their research, so central to their credentials, and the
apparent purpose of *Forensis*, whose subtitle was, after all, 'Towards an Investigative Aesthetics,' is put on display. These questions are given another dimension
when we take into account how Forensic Architecture's work is often offered
as evidence by lawyers representing the families of victims of police and state
brutality, and enrolled or even commissioned by organisations such as European
Human Rights Advocacy Centre, Amnesty International, Human Rights Watch
and Médecins Sans Frontières.

Embedded within FA's exhibition my experience was of being a viewer and
a reader simultaneously. This was not really an exhibition in which objects were
the focus and text was a caption. Instead, the sheer volume of words gave them a
force and meant that as a visitor I felt compelled to spend more time with them

than I might do with a standard gallery information board. The text announced its presence in other ways too: the galleries at times seemed ill-configured to contain it; the seating was in the wrong place; the flow of people, organisation of works, and so on seemed to create congestion, less around certain works – as is often the case in front of famous paintings for instance – and more around volumes of text. In terms of content, the exhibition had the feeling of a series of case-studies, specific projects, a litany of horrific human rights abuses, and terrifyingly detailed war crimes. In each case, the text offered as much context for the project, as it did a detailing of the methods through which the team developed their research. As a result, as I explored the exhibition further, the projects while powerful in their own right, when seen together created a powerful sense of FA's overall research practice.

Perhaps unsurprisingly a particular account of research came into view, foregrounding the creative aspects of their practices. This included a detailed video of the use of computer-aided design software to work with ex-detainees to model the spaces in which they were held. This process involved a sort of visually prompted interview in which the ex-detainees where asked detailed questions about their sensory experiences of the spaces they were held in. This included, for example, exploring the shifting pattern of sunlight, the number of guard's footsteps down the corridor, or the height of the slots in the cell doors that food was pushed through. This information was then used by the computer operator to model this space on the screen in front of them, making tweaks and adding details as the picture was built up through further memories prompted by the model emerging on the screen. This display of working process was common throughout the exhibition; whether a detailed video, the visualisation of case study development from crowd-sourced social media and radio reports, or the display of a typology of aerial photographs sourced from multiple locations – drones, weather satellites, and so on. The practice of looking as an audience member becomes an apprenticeship of sorts into FA's methods. Encountering the image sequences, became a practice of getting your eye in, of being guided by their accounts towards a noting of minute differences, such as the shifts in the profile of a building through its shadows, or gaining the means to read across multiple difference visual sources to mentally compile before and after pictures. I felt the exhibition was not just telling me about FA's investigative aesthetics, but beginning to train me as an investigative researcher.

The exhibition was hugely successful at sending me away with a sense of these creative and aesthetic elements of FA's research process, and also how these intersected with more scientific forms of evidencing – ballistic testing, for example. What, however, I was less clear about were the more standard social science research practices, such as interviews or ethnography, which were clearly an important part of their research practice. Nor did I get a sense of what must be painstaking practices of secondary analysis of text and image, or the means and logics by which the images and films were compiled and edited. Also missing, surprisingly, was any sense of the ethical procedures of these investigations. There was little indication of the degree of ethical complexity that these researchers must have encountered. In the reconstruction of the spaces of the Saydnaya Prison operated by the Syrian government, we watch people gathered around a computer screen effectively requiring the sharing of vivid memories of torture to materialise in front of us a model of these clandestine carceral geographies. We are invited to consume this process; to celebrate the ingenuity and potential of combining sensory experience and computer-aided design technology. Yet at no point did the video of process suggest how concerns with re-traumatising victims had been managed or how the team were engaging with practices of anonymity and ensuring the safeguarding of the participants. I am not suggesting that the FA team did not attend to these aspects in their research process, I am sure they did, and indeed they would have had to as a condition of the various forms of research funding they received. Yet it was notable that these aspects of their process were not considered important for display. I was left with a series of questions about their investigative aesthetics, aware of certain innovative aspects of their research practice, but less clear about some of those more mundane but vital elements. Loosening the public accounts of their research process from some of these aspects might come at a cost. For, if text can 'author' how we look, as the now caricatured post-modern 'death of the author' reminds us, it cannot control how we look, however hard we might try.

Exhibitions are always, as Felix Driver wrote of his experience of curating an exhibition of objects from the Royal Geographical Society's collection, 'inevitably a work of transformation.'[41] Driver muses on the effects of displaying archival materials, and the tension between the researcher who wants to remain true to the materiality of the collections, their imperfections, their role within the wider archive, and their desire to use the material to present a powerful story.[42] He describes it as a case of,

> juggling two different approaches to the use of historical materials in the display space. The first approach gave priority to principals of archival authority, the need to display materials in or near their original form, either as objects or as faithful reproductions, the second sought to align the spatial form of display with the intended message or ethos of the exhibition.[43]

The experience of FA presents us with a cautionary tale about the transformatory possibilities of exhibitions. For their aesthetic evidence has not just formed art 'exhibits' in art galleries, but also 'exhibits' in court rooms around the world, and this dual presentation has not sat easily together.[44] As Eyal Weizmann, the group's leader, observes in an interview with leading art journal *October*,

> Human rights groups traditionally relied on art to add affect to the investigative process, and implied that the investigative process is too serious for aesthetic practitioners. The court itself is allergic to the work of aesthetics and art because it sees in them the danger of manipulation, emotional or illusionary, that takes the viewer aware from supposedly unmediated experience.[45]

Interestingly, behind these discussions lingers a sense that some 'data' put on show is objective, but becomes later tainted through its alignment with art, reduced to spectacle, challenging the 'critical value of work that has at its heart a form of justice seeking.'[46] Indeed as a member of FA noted 'the new artist status may even weaken FA's investigations'; pro-Assad Groups have sought to dismiss their work with Amnesty International exposing Syrian torture: '[they] accused us of being artists, of it being special effects, of fabricating the evidence.' While the Christian Democratic Party has described them as an 'unserious artist group.'[47] In response, the group have not sought to debate the place of art as 'evidence,' but have instead sought to asset the role and presence of forms of expert testimony from 'ballistic, fluid dynamics and acoustic experts' within and backing up their cases.[48]

FA is perhaps an extreme example. The research agency is situated in the midst of a series of emotionally and politically powerful issues, often working on cases with critical urgency and they evolve working methods whose very title 'forensic aesthetics' mobilises the tension between arts and science. Yet, they bring to the fore important questions around how, in the midst of excitement over the possibilities of research as exhibition, we address the use of research data and findings as objects for/of display. We are reminded how exhibitions, just as much as articles, are processes of the framing and transformation of data. Whilst geographers might be familiar with these practices in articles and have a series of careful protocols around anonymity, around meta-data preservation and storage, around redaction and anonymity, we are less experienced at doing this within exhibitions, and in thinking through the implications of the transformation enacted by the exhibitionary context. For some research whose value is based in epistemologies of fact and truth, where its objectivity is integral to its wider value as evidence or testimony, the exhibitionary context might enhance this value, but it might also undermine it. For while, without question, exhibitions can enable the presentation of work to new audiences and the raising research profile, they can also do damage. The exhibitionary context (with its associations with artistry) can offer those seeking to undermine the value of evidence unpalatable to them, a means through which to challenge the data's objectivity, thus risking undermining the important work that these materials might do. Of course, most 'research on display' is far less embedded in the art world than FA's, and has far less budget and focus on exhibitionary practice. Yet researchers putting data on display still have to grapple with questions around the transformations of that data, around the display of research process and the framing of findings. In what follows I want to explore how far from just transforming research, exhibitions might themselves come to constitute research.

Exhibiting research 3: exhibition as a 'Research Action'

Basecamp, the project was called. It was an exhibition of a different form. In a 'meanwhile' space, part of an old school used as an arts venue while the developer was raising money to convert it into a block of flats, each exhibition somehow – however lightly – we reflect, complicit in gentrification.

Rather than a finished set of objects, this exhibition as base camp was a point of departure for an exploration, a point of beginnings, of questions and possibilities, rather than a presenting of results. When you entered the room and saw a map of an ice cave, "start here," there was a tea stop, an abyss, a series of stations, activities, and ideas. As 'audience' you circulated through the space, conducting the different activities, drawing around your body, arranging text, interacting with objects, and attempting to transmit your own thoughts to others. I sat in the corner, my research note-taking on view as part of the process of the exhibition. During the three days I worked as part of this project, I watched as Flora and Kelly planned out the space as a kind of investigative game, exploring a series of instructions for navigating stations that visitors could travel through and at each stop doing a series of activities. The exhibition space felt more like studio space or workshop, ideas in motion rather than the 'display' of something already complete – a learning process which was fed by the active role of the participants. Parts of previous works were present, recycled into this work, other elements continued forward into other works, new ones were created over the course of this exhibition, some fell away.

From field-notes, August 2017

Writing in 2014, Paul O'Neill and Mick Wilson identified two modes of the interaction of research and curation, 'researching within exhibition-making' and 'exhibition as a research action itself.' Simon Sheikh, extending this observation reflected,

> the curatorial project – including its most dominant form, the exhibition - should thus not only be thought of as a form of mediation of research but also as a site for carrying out this research, as a place for enacted research. Research here is not only that which comes before realisation, but also that which is realised throughout actualisations.

This is akin to the increasing common question, what does it mean to curate in a non-propositional way? In other words, where the exhibition does not seek to be a presentation of findings or a final statement on a topic, but rather where the relations between curator, exhibition, audience, and social context are an active site of knowledge production?[49] Interestingly, the resistance to the exhibition as end point is more diffuse than we might think. Indeed, historical geographer Felix Driver described his experiences as less a putting of research on display and more as part of the process of shaping research.[50] Reflecting on the exhibition 'Hidden Histories' he and Lowrie Jones evolved with and at the Royal Geographical Society in London he ponders,

> the relationship between research and display was by no means all one-way, as the language of "dissemination" tends to suggest. The process of bringing the exhibition into being- conceptually, discursively, and

practically- also helped to reshape research questions and perspectives in ways that were productive of new insights about the subject of the research, and the process of public engagement.[51]

Driver discusses the curatorial experience of the exhibition, reflecting on how the design of the experience across both the street level exhibition space – the Pavilion Gallery and the Foyle Reading Room below (linked to the library and archive) – was both a function of conservation, but was also a reflection of the desire on the part of the curators to link 'stories to sources.'[52] The upper level space included panels of text, prints, and film and some artefacts, whilst the space below was focussed on oil paintings, books, sketches, and artefacts. This division of attention across these spaces embedded the links between 'exhibition and research in the spatial organisation of the display.'[53] Further, it was suggested of the audience that 'in crossing the threshold of a formerly inaccessible research facility they were invited to become active participants in the making of new knowledge rather than simply its passive spectators.'[54] Taking up this question of how the research exhibition might be enfolded with the research process rather than come after it, I want to explore how a series of exhibitions and events I was involved in might best be thought of as 'research in action.'

As previous chapters in this volume have noted, some of the key questions art historians have asked of shifting practices of display and public engagement – whether this be live art, event-based work, or participatory practices – concerns the kind of audience experience it convenes and the sets of material practices that have come to constitute exhibition-making.[55] In other words, is this audience as spectator, as viewer, as participant, or in this case as co-researcher? Reflecting on *Basecamp* it seemed that what was 'on- display,' was not a series of objects to be consumed, information to be examined and processed, but rather, a set of activities to be participated within, thoughts and reflections to be stimulated and created through the experiences. There are models for thinking through the research exhibition in this way. Sociologist Bruno Latour, for example, has been experimenting with the exhibition as a site for thought-in-motion, rather than as a local for already thought and presented ideas. He writes of his exhibition *Reset Modernity* (2016):

> I call this show a Gedankensausstellung, thought exhibition, in the same way as people talk about Gedankenexperiment [thought experiment]. It is a gedankenausstellung, in the sense that it tries to present a problem. It's a conceptual point: can we think of politics in other terms than usual ones, by turning to things? What would politics turn into without centring in human opinions? But it's a show anyway. People will come for fun in here… the exhibition is not spectacular. It has very few works of art to display. But neither is it an exhibition about documents in cases for people to read. It is more like a hands on Science Museum/Lab where visitors can feel and test for themselves what sort of contradictory accounts they have

of 'modernity' this most difficulty term to define. ... visitors... are transformed into co-inquirers.[56]

Bruno Latour's extensive website records his role in exhibition making and enables us to see how he situates it within his wider disciplinary practice. His curatorial experience includes three large-scale exhibitions at the Centre for Art
and Media in Karlsruhe (ZKM): *Iconoclash: Beyond the Image Wars in Science Re-
ligion and Art* (2002), *Making Things Public: The Atmospheres of Democracy* (2005),
and *Reset Modernity!* (2016) as well as a play, several performative re-enactments,
and numerous curatorial interventions, such as the *Monument to the Anthropocene*
(developed in collaboration with sociologist Bron Szyzersinksi).[57] His writings
have been much beloved by art theorists and practitioners, and in 2010 he was
awarded the Nam June Paik Art Centre Prize.[58] He has also co-founded an
experimental research programme in art and politics, *Sciences Po École des Arts
Politiques*, based in exploring common modes of enquiry across different disciplines.[59] From in the midst of this work, Latour's writing and practices offer us a
model of an exhibition as an alternative public site. The exhibition emerges less
as a power-knowledge nexus where the institutional gaze disciplines the bodies
and minds of its visitors, and rather frames exhibitions as sites for question and
query. As such, visitors are co-inquirers, 'feeling for themselves' the problems the
exhibition frames. The exhibition is thus less a vehicle for displaying evidence or
even the research process and rather emerges as a conceptual form based around
a series of investigative operations. In short, the exhibition could be considered a
research action. To investigate this further I want to turn to a final example of a
research exhibition: *The Artificial Cave*, held in the David Roberts Foundation,
London, in summer 2016.[60]

*We stand around in the gallery space, assessing the fibreglass sections, the blocks of
wood, the box of tools, the slightly dirty floor of the East London project space. We grasp
one of the sections, its heavier than I thought, and awkward, bulky with some sharp edges.
Someone gets some bolts, wing-nuts, and a wrench. We start to assemble the cave. There
are straight bits, tight corner sections, more relaxed curves, sometimes the rough black outside
does not entirely mirror the nature of the interior topography of these short tubular sections.
In places, the chicken wire that forms the structure has rubbed through, elsewhere the surfaces have been worn smooth by all the bodies that have squirmed through this artificial
landscape (Figure 8.6).*

*We have gathered as a group to explore this cave, to use this artificial cave as it comes to
be called – a training cave it actually is, as a sort of prompt to thought, perhaps to draw out
differences and similarities between this caving and the caving we had done in the landscape
several weeks earlier and several hundred miles away. Could we replay with our bodies,
remake using this cave or just on the floor, the kinds of journeys we had gone through underground. We squiggle and squirm our way through this space, we reflect on its materiality,
its lack of darkness, the sense of safety and the known we experience, we talk about what
this cave 'lacks' in relation to that cave, eventually we turn to what it opens us onto. Later,
a sort of symposium is held around the artificial cave, people offer up artworks, scientific*

FIGURE 8.6 *The Artificial Cave*, David Roberts Foundation, London, 2016. (Image, author's own)

models, collages of text, other experiences of extreme sports. None of these have been created in relation to that fibreglass form, but all are now explored alongside it..

From field-notes made after The Artificial Cave, 2016

The *Artificial Cave* was an exhibition in a somewhat ambiguous way. Held in a gallery space, on display for only a few days and only by invitation, the gallery space became a site for creative collaborative thinking connected through responses to the fibre-glass form in front of us. The gallery in this case offered the chance to stage a space, to create, what to return to Chapter 1 on field, might be thought of as a 'thinking space.' We came together to explore the artificial cave, drawn by it, and other props, rocks from the original caving expedition, images, shared and individual memories, conflicting and interweaving. This was a collective thinking process, whose similarities and differences gave rise to 'new thoughts through interaction.'

In an earlier workshop with Flora working with PhD students exploring ideas of practice-based research in geography, Flora had introduced us to the Brazilian artist Lygia Clark (1920–1988) as part of a workshop activity. During this activity we donned black t-shirts bought for the occasion, and then with large needles and coloured twine, we followed a series of movement-based instructions – turn to your right and move up and out – that resulted in us all twisted and stitched together in a multi-limbed relational field. Such contact was challenging for many. The experience had lodged in my mind; unsure what to make of it at the time, I continued to ruminate on it. It was odd to come across Lygia Clarke again several years later, this time as a figure inspiring Erin Manning's understanding of practices of research creation and the lure of art.[61] I checked with Flora, she did not know of Manning's work, this was largely beside the point. Although interestingly, if Manning and SenseLab had been used to frame the workshop activity, it might have offered a hook people could accept to frame, if not understand the embodied activity and the relational field we created through it. It was interesting that Flora encouraged us to feel and experience the effects of Clark's activities, whereas Manning's text, even as it tries to avoid doing so, explains Clark's work by reference to theoretical figures, Whitehead, Deleuze and others. Manning uses Clark's work – air, rocks, bags of sand – weird ugly forms to displace the focus on the object and its representational possibilities towards the artfulness of the 'lure' of these objects, their blocs of sensation and provocations. Clark's 'relational objects' were not 'valuable' or 'artistic' abstracted from what they could do, they were 'just' bags, 'just' rocks, but what was important was the 'relational field they actuated', their artful ness was in the environment they co-created. To return to our artificial-cave on display, ugly, bumpy, not-scary (like the real cave), slightly smelly, plastic not millennia old rocks.[62] This was an odd lure, not a romantic lure, or the sublime force of other caves, but rather a provisional place that we had created amongst us. Our encounters with it drew us apart and together in different ways. We collaborated to create its form, to crawl through it, to take the rock through it, to reorder the sections, to direct each other. Its artificial rocky-body, our bodies, emerged differently each time; in each interaction we never knew what would happen. What does this somewhat abstract experience mean for the idea of a research exhibition? This plastic artificial cave, the collective of people it lures, the relational field it creates, across and between us, this cave and the ones before it, together suggest the exhibition might be a thinking-space, a space in which objects and engaging with objects is thought-in-action, rather than a display of research already thought.

Three very different exhibitions, *Basecamp, Reset Modernity*, and the *Artificial Cave*, between them demonstrate another iteration of the research exhibition, in which to explore research is to become enrolled in part of the process, to be led to think, rather than told about things to reflect on. To engage in activities, to be part of a thought experiment, with no predetermined outcome but rather as a space in which to inspire the audience to active engagement and questioning. In different ways these three examples situate the exhibition as part of an ongoing research process, not just for the visitor but also for the curator. Importantly though, whilst Latour seems to maintain a certain kind of intentionality and directionality, he knows the kinds of audiences he wants to assemble. In *Base Camp*, there was less clarity over the kinds of thinking and questioning that the exhibition wanted to bring about, but there were constraints offered through the activities and the instructions provided. The intention was as if to enrol the audiences in the ongoing process of the work and the research, where audience participation and active debate evolves the exhibition from moment to moment and also from where, as Kelly and Flora discuss, the next interaction emerges. Krysa makes a similar observation of contemporary exhibitionary practice, observing,

> situating exhibitionary practices at the intersection of academic research and public display thus expands the traditional remit of a gallery – as the privileged site for staging exhibitions or pedagogical resource – to the idea of a 'lab' where experimental thinking and making can take place.[63]

The process of curating a research exhibition might less be about object making, and rather about creating a thinking space, a setting of the conditions and frameworks for experimentation and the dynamic process of asking questions with audiences in exhibition spaces – this is curation in a non-propositional form. The exhibition is therefore not the result of the curatorial or research process, but instead indicates that curation is research in itself, where 'questions are not answered by recombined in the very act of making.'[64]

The geographical research exhibition?

For a long time, galleries and museums, especially large institutional spaces, had long been conceived of by the avant-garde as akin to cemeteries or graveyards for critical thinking and ideas. Yet for some these spaces offer the chance to evolve sophisticated critiques of the representational challenges of gallery's and museum spaces as well as for other wider concerns, not least those of geographical research. This chapter has focussed on the evolving form of the geographical research exhibition and has queried the range of forms this might take, from putting data and artefacts on display, to installing elements of the research process, to the exhibition as a more propositional form encouraging the audience to be a co-researcher. As with previous discussions, the point is not to suggest a singular correct form of research exhibition but rather to explore how it is that curatorial practices have come to be both a site of intersection of geographical research and creative practices, whether in terms of geographer-as-curator, or whether that be the curation of work produced at the intersection of geographical research and creative practices. What is intriguing is that many of the same issues that have been negotiated throughout the volume emerge here, around the nature of 'data'; if one does not extract 'data' from a field site then it is perhaps hard to put data on-display; for some it is the process, the doing, that is what is put on display and that is the point of the practice. Though as with FA's exhibition there is clearly much at stake in displaying your process, for what you put on display, is transformed by that context, sometimes into something more artful and thus less apparently objective than one might wish. Further, there are clear questions about what one chooses to display or not, whether that is images that might feel intrusive or dangerous to those they picture, or whether that is elements of a research process that are perhaps not as interesting, innovative, or aesthetically pleasing, and driven, as, say, computer-aided design reconstruction. How for example, might one display ethics processes, how could one put on display safeguarding practices and data storage? Such exhibitions of research create different experiences of their audiences. Exploring FA's processes-based display, I had to read a lot, more than I would normally expect to, it was like being in a visual essay, but yet, by the end I felt as if the image-based works combined with the text had effectively tutored me in their ways of working, a powerful way to turn exhibition goer into researcher. Yet there was a distinction between FA's enrolment of the exhibition audience and *Basecamp* and *The Artificial Cave's* forwarding of the experience of the audience as that of co-researcher. Now clearly some might have been alienated by the activities, or found it hard to participate in the encounter with the fiberglass cave. Yet it felt that across these exhibition-events the research exhibition emerged as a propositional form. In other words, less a definitive statement about something and more an invitation for those attending to explore, to engage with the spaces, to think them through – exhibition then as research action. As such, the exhibition can no longer be situated as a form of output, one in a series of linear, successive frames an artwork or artefact moves through – field, studio, gallery and so on, rather, as other chapters have suggested these spaces should be considered as intertwined, overlapping practices of the production, consumption, and circulation of geographical knowledge. It seems to me that if exhibitions are going to be reclaimed, as many wish as spaces for experimentation, then conceiving of the exhibition as a site of geographical research in action might be one interesting way to think of the potential of the geographical research exhibition.[65]

Notes

1 Ferguson et al., "Mapping International Exhibitions," p. 2.
2 Bennett, "Exhibitionary Complex."
3 See for example, Driver, *Geography Militant*; Donadelli et al., "Introducing the Padua Museum of Geography."
4 Ryan, *Picturing Empire*; Hayes, "Slidescapes"; Neidderer et al., "The Research Exhibition."
5 O'Neill and Wilson, *Curating Research*.
6 Voorhies, *Beyond Objecthood*.
7 Rust and Robertson, "Show or Tell."
8 Neidderer et al., "The Research Exhibition."
9 https://www.tate.org.uk/context-comment/audio/beyond-academy-research-exhibition-symposium-audio-recordings (accessed 23/12/2019).
10 https://www.kew.org/science/our-science/people/felix-driver (accessed 23/12/2019); https://www.neonmuseum.org/newsletters/march-2019/scholar-in-residence-lecture-untold-history-of-neon (accessed 23/12/2019); https://www.terrain.org/2014/currents/biosphere-2-poetry-anthropocene-eric-magrane/ (accessed 23/12/2019); https://www.capefarewell.com/2011expedition/author/libby-straughan/(accessed23/12/2019);https://bundanon.com.au/residencies/2018-residencies-list/ (accessed 23/12/2019).
11 For example, Dr David Rooney, Keeper of Technologies and Engineering at the Science Museum, London and Curator of Timekeeping, Royal Observatory, Greenwich; Dr Phil Hatfield, Head of the Eccles Centre for American Studies, British

Library, London; Dr Ellie Miles, Documentary Curator, London Transport Museum; Dr Bergit Arends, Collection Care Research, Tate.

12 See for example, Cecilie Sachs-Olsen, curator of the Oslo Architecture Triennale (2019), http://oslotriennale.no/en/aboutoat2019 (accessed 23/12/2019) and Sasha Engelmann, who curated a series of public events for Studio Saraceno during a major retrospective of the studio's work at the Palais de Tokyo. https://www.palaisdetokyo. com/en/event/voices-collide-day-aerocene (accessed 23/12/2019).

13 See for example, Stephen Daniels has co-curated a series of major exhibitions at key international arts venues, including, The Art of the Garden (Tate Britain, 2004), https:// www.tate.org.uk/whats-on/tate-britain/exhibition/art-garden (accessed 23/12/2019); Paul Sandby: Picturing Britain (Royal Academy, 2010), https://www.royalacademy. org.uk/art-artists/archive/paul-sandby-ra-picturing-britain (accessed 23/12/2019); http://www.artistsinlabs.ch/en/research_publications, (accessed 23/12/2019); https:// walthamstowwetlands.com/fearghus-oconchuir, (accessed 23/12/2019).

14 See for example, Sarah Hall's Everyday Austerity exhibition, https://everydayausterity.wordpress.com/exhibition/ (accessed 23/12/2019).

15 See for example, Driver et al., "Landing"; Karen Till, Land~Edge, The Glucksmann Gallery, University of Cork, https://maynoothgeography.wordpress.com/2017/05/02/ maynooth-at-the-conference-of-irish-geographers/ (accessed23/12/2019); MerlePatchett and Neville Gabie, https://merlepatchett.wordpress.com/2013/05/20/archiving-oil-images/ (accessed 23/12/2019); Donadelli et al., "Introducing the Padua Museum of Geography."

16 See for example, http://www.mappingspectraltraces.org (accessed 23/12/2019).

17 See for example, Karen Till, Land~Edge, The Glucksmann Gallery, University of Cork, https://maynoothgeography.wordpress.com/2017/05/02/maynooth-at-the-conference-of-irish-geographers/ (accessed 23/12/2019).

18 https://www.theguardian.com/global-development/gallery/2018/oct/18/how-climate-change-forces-farmers-into-slavery-in-cambodias-brick-kilns-in-pictures; https:// www.bisonbison.co.uk/work (accessed 23/12/2019).

19 https://www.omidyar.com, Omidyar is known for working with George Soros to create fact-based News services (accessed 23/12/2019).

20 https://www.societyandspace.org/articles/blood-bricks-untold-stories-of-modern-slavery-and-climate-change-from-cambodia (accessed 23/12/2019).

21 https://www.projectbloodbricks.org/project (accessed 23/12/2019).

22 https://www.projectbloodbricks.org/events/2018/7/2/global-challenges-creative-turn-a-workshop-exploring-the-possibilities-and-practicalities-of-creative-research-methods-and-practices-in-the-global-south (accessed 23/12/2019).

23 https://twitter.com/hashtag/livedausterity (accessed 23/12/2019).

24 https://www.societyandspace.org/articles/blood-bricks-untold-stories-of-modern-slavery-and-climate-change-from-cambodia (accessed 23/12/2019).

25 Vergo, "The Reticent Object."

26 https://www.magnumphotos.com/about-magnum/overview/ (accessed 23/12/2019).

27 The Guardian (18/10/18). How climate change forces farmers into slavery in Cambodia's brick kilns – in pictures. By Hannah Summers and Thomas Cristofelleti. https:// www.theguardian.com/global-development/gallery/2018/oct/18/how-climate-change-forces-farmers-into-slavery-in-cambodias-brick-kilns-in-pictures (accessed 23/12/2019).

28 Bishop, "Information Overload."

29 Ibid.

30 Ibid.

31 Interview notes.

32 https://www.rgs.org/getattachment/Research/Journals,-books-and-guides/RGS-IBG-Book-Series/RGSCommunicatingGuide2013.pdf/?lang=en-GB and also QMUL's guide for the production of exhibitions, https://www.qmul.ac.uk/geog/media/geography/docs/staff/Research-on-Display-Final.pdf (accessed 23/12/2019).

33 Gat, "Could Reading be looking."

34 Ibid., p. 4.

35 https://forensic-architecture.org (accessed 23/12/2019); Weizman, *Forensic Architecture.*

36 For an archive of past and forthcoming exhibitions please see, https://forensic-architecture.org/programme/exhibitions (accessed 23/12/2019).

37 https://www.macba.cat/en/exhibition-forensic-architecture; https://www.tate.org.uk/whats-on/tate-britain/exhibition/turner-prize-2018/forensic-architecture; https://www.whitney.org/exhibitions/2019-Biennial?section=22#exhibition-artworks (accessed 23/12/2019).

38 https://frieze.com/article/id-rather-lose-prizes-and-win-cases-interview-eyal-weizman-turner-prize-nominated-forensic (accessed 23/12/2019); https://www.thetimes.co.uk/article/turner-prize-hopefuls-we-re-not-artists-xtq7c7knq (accessed 23/12/2019).

39 https://www.dezeen.com/2018/05/04/forensic-architecture-turner-prize-warning-phineas-harper/ (accessed 23/12/2019).

40 https://www.thetimes.co.uk/article/turner-prize-hopefuls-we-re-not-artists-xtq7c7knq (accessed 23/12/2019).

41 Driver, "Hidden Histories Made Visible."

42 Ibid.

43 Driver, "Hidden Histories Made Visible," p. 432.

44 https://frieze.com/article/id-rather-lose-prizes-and-win-cases-interview-eyal-weizman-turner-prize-nominated-forensic_(accessed 23/12/2019).

45 Bois et al., "On Forensic Architecture," p. 122.

46 https://frieze.com/article/id-rather-lose-prizes-and-win-cases-interview-eyal-weizman-turner-prize-nominated-forensic (accessed 23/12/2019).

47 https://frieze.com/article/id-rather-lose-prizes-and-win-cases-interview-eyal-weizman-turner-prize-nominated-forensic (accessed 23/12/2019).

48 https://frieze.com/article/id-rather-lose-prizes-and-win-cases-interview-eyal-weizman-turner-prize-nominated-forensic (accessed 23/12/2019).

49 Sheikh, "Towards the Exhibition as Research"; Krysa, "Exhibitionary Practices."

50 Driver and Jones, *Hidden Histories.*

51 Ibid., p. 86.

52 Ibid.

53 Ibid.

54 Ibid., p. 88.

55 Bishop, *Artificial Hells.*

56 Latour, *Reset Modernity.*

57 http://www.bruno-latour.fr/node/582.html; https://zkm.de/en/person/bruno-latour (accessed 23/12/2019).

58 https://www.e-flux.com/announcements/36186/bruno-latour-wins-nam-june-paik-art-center-prize-2010/ (accessed 23/12/2019).

59 https://www.artandeducation.net/schoolwatch/58047/reassembling-art-pedagogy-pragmatism-inquiry-and-climate-change-at-sciencespo-experimentation-in-arts-and-politics (accessed 23/12/2019).

60 Hawkins, "Underground Imaginations."

61 Manning and Massumi, *Thought in the Act.*

62 Hawkins, "Underground Imaginations."

63 Krysa, "Exhibitionary Practices."

64 Ibid., p. 69.

65 Raunig, "Factories of Knowledge."

CLOSING THOUGHTS

'Lipstick on a pig' was how one usefully honest, social scientist friend of mine responded to my practice attempt at justifying my research approach to an interview panel. This was ahead of the interview that opened the introduction to this volume. By this, I think he meant, supportively, that I might want to rethink the way I was positioning my research. For while, try as I might to situate artistic practice within geographical knowledge in the context of social science terms and discussion, to justify empirical approaches that enabled an accounting for senses, emotions, and affect, and even to offer points of 'evaluation,' it was not working. From his perspective it felt like a superficial attempt to hide the nature of what I was doing: lipstick on a pig. He was right. The alternative was to acknowledge the differences, that some of this work at the intersection of geographical research and creative practices can be aligned with understandings from across the diverse epistemological, ontological, and methodological field that is social science, but that much of it can't. To pretend that it can be would be antithetical to its premise. For alongside research that situates these approaches in the context of certain understandings of the empirical that stem from positivism and social science more generally, we also need research that embraces different approaches and celebrates their possibilities. Further, however, we also need not just to embrace different approaches, but to also ask critical and sometimes difficult questions of these approaches. It is such work that this volume has tried to do – to explore the production, consumption, and circulation of work produced at the diverse intersections of creative practices and geographical research but also to probe this work critically. I have sought to conduct this 'research on research' in a way that is suitably chastened by the concerns expressed by those working across geography and across practice-based research, that the creative turn and the research turn are in need of further critical thinking and engagement. I have tried hard not to get carried away with the fetish of creativity. My aim here has not been to offer a guide to best practice, to review the field, or to summarise or canonise certain pieces of work, individuals, or ways of working; instead, I have tried to explore some of what it means to actually do research at the intersection of geography and creative practices. And from this position 'in-the-midst' to reflect on wider questions of the production, consumption, and circulation of this research.

I have framed this working 'in the midst' as a paraethnography, a research-ing of research, which sometimes felt like a risk and oftentimes felt like a fraud. I have also framed this in terms of an experiment with the geohumanities. In doing so, I understand the intersection of geography and the arts and humanities that the geohumanities foregrounds as appropriate response to both the nature of this field, and what I consider to be its 'forgetting' of both geography and the humanities as it has sought to gain critical perspective on itself. At a time when those across geography and the arts and humanities were urging a greater criti-cality for these ways of working, I wanted to respond with a form of criticality that met these approaches to geography and creative practice on their own terms, for me this involved exploring what an admixture of work from geography and the humanities might offer.

It felt that this diverse field of geographical research and creative practices, the rich range of ways of producing research and the forms it took, demanded an equally rich response. As such, the volume has drawn from across art theory and history as well as histories of geography. The risk, I am aware is that it ap-pears both too much like review and also too superficial. Yet the form of the book was also inspired by work from across those fields. This includes those art historians and theorists who have offered expansive accounts of the emergence of forms and understandings of art across time. Whether that be the demise of the Renaissance artistic team, and the emergence of the Romantic image of the individual artistic genus. Whether that be the ways live or performance art, con-sumed on the spot by audiences and also documented in myriad innovative ways, shifted understandings of how we consume art and how it can circulate away from gallery and dealer systems. Whether that be the ways that participatory art demanded we rethink how it is we judge art, through aesthetics or a more soci-ological frame, or through a reworked understanding of both? Such accounts are often expansive in the scope, moving across a series of questions of the art's pro-duction, consumption, and circulation by means of a diverse series of examples. This has clear resonance with this volume's approach. Inspiration also emerged from those historians of geography and of science, who have asserted the need to 'put science in its place' and sought to map various sites and scales of the pro-duction, consumption, and circulation of geographical knowledge.[1] The third set of inspirations that have shaped the form and querying of this volume have been those accounts of working across, within, between, and amidst disciplines. Whilst many work in these ways, and there is a wealth of literature on terms to use, there are still relatively few attempts to account for these ways of working in practice. As those working at the intersection of creative practice and research from beyond geography have noted, we have still failed to address how 'the process of art and creative practice alter[s] what we might think of as research?'[2] Further, we need, it is suggested to take more seriously questions around how we define disciplines in this context, 'how it [a discipline] might be defined by how it escapes its past content and internal constraints... how it renews itself through the rigorous indiscipline of its effective couplings with processes other than its own.'[3] Working with these inspirations from in the midst of this field, I want to offer four sets of thoughts to close. The first three of these address each of the volume's three trajectories in turn; the fourth reflects on the geohumanities experiment of the volume, and what lessons this might offer for the wider progress of this field. In the spirit of my attempts to reflect from in the midst of a field of research that is still on folding, these are more in the tenor of thoughts and reflections than any form of 'conclusion' as such.

The first of the book's trajectories was to explore some of the key sites of the coming together of geographical research and creative practices. As such, the backbone of this text has been formed by the organisational logics of eight site-based ethnographies that have explored some of the more 'charismatic' sites of the relationships between geographical research and creative practices – the field, the studio, the exhibition, and so on. Whilst geographer's engagement with the artist residency might be a relatively new occurrence within the discipline, the studio and the exhibition have long offered, as was hopefully indicated, a site for the coming together of geographical knowledge and creative practices, as indeed has the field. There are other sites that, of course, could have been explored here – the body, the archive, the 'alternative' art institution – as well as sites, like the conference or the seminar. Indeed, some of these emerged over the course of discussions, others fell by the wayside. There are still others, such as the institution, that have emerged obliquely through discussion rather than as the focussed subject of discussion.

Just as we now recognise the partiality of all mappings, this volume has not sought for an exhaustive completeness, and as this is not a gazetteer it does not aim at the equivalent of a 'Capes and Bays' account of the field, and it has perhaps not been as literally geographical as some might wish. In other words, this volume did not seek to take stock of the global patterns in the uptake of this research, although, as noted in the introduction, at the time of writing we could find these coming together in countries around the world. Instead this is a 'mapping' that under the influence of those geographical accounts of knowledge has sought, through its grounded attention to a specific set of sites, to direct our attention to the need to think across sites and practices of the production, consumption, and circulation of these ways of working. As chapters like the thesis, exhibition, and page demonstrated, not only is each a site of production and consumption, but these activities are thoroughly implicated. When geographical research takes form in installation, performances, films, and events, we should also consider, how it can find a place on the page, how can it be accounted for in the context of research cultures that foreground so-called linguistic imperialism, and the ways that permeates practices like citation as well as points of judgement: thesis examination, job interviews, and promotion? As the intersection of projects and sites across the chapters has hopefully also made clear, the relationship between the geographies of production, consumption, and circulation must also not be conceived of in a linear manner. For, as artist Daniel Buren noted, we need to disrupt any sense of a neat series of successful frames of progress of an art-work from the studio, to the gallery and thence into catalogues, magazines, and books.[4] In the place of linear spatial progress we need to appreciate a non-linear and relational account that also expands the spaces it considers to be those of knowledge production at the intersection of geographical research and creative practices. To appreciate this is to be directed towards a clear sense that, as with the histories of geographical knowledge, these forms of knowledge making re-quire we consider relations – connections and disconnections, and the mobilities and immobilities – of these forms and knowledge production, their methods, and their outputs. Another dimension of these mappings that emerges clearly is that of scale. In other words, just as geographers studying the geographies of the book remind us to look up from the page to consider factors like national and inter-national review cultures, as well as the knowledge circuits created by Empire, so we need to do the same here. We need to consider not just the geographies of exploring the individual creative practice in front of us, but step back from the sculptural form or remove our ear-phones to also consider national and interna-tional cultures of review and evaluation, within and beyond our university or art world institutions and our various funding bodies.

The very situated accounts that the eight chapters have offered of particular projects that intersect geographical research and creative practice have sought to echo the commitment shown by historians of geography (amongst others) of late to small stories, everyday practices and the individual and collective doings of ordinary academics. To do so is to recognise the very practiced nature of disci-plines and institutions and knowledge making. It is also to try to appreciate the value that grounding an intellectual field might offer as both a way to view past histories but also as a means to appreciate and sit with the context of a field of work very much in motion. To speak of mappings, then, is less to evoke a sense of the power-knowledge of the artefact 'map' than it is to learn from those for whom mappings – as a practice –shapes knowledge and imaginations but might also be conceived of as always ongoing and provisional.

In exploring these sites the volume's second trajectory was concerned to track and trace some of the ideas of research, and linked to this geography and creative practices that emerge. Being committed to the ways of working in the midst of this field also meant guarding against the production of a singular account. As an example, the chapter on the field considers multiple forms of creative fieldwork-ing that occurred just over the course of one project and tries to grapple with the conflicting epistemological traditions to which they might belong and through which they might be understood. A similar sense of multiplicity permeates dis-cussions of the research exhibition: sometimes this is about the display of data, at other-times the exhibition is a proposition, a provisional form that casts the audience as co-researchers. The chapter on the lab opens out these debates in a somewhat different way, as it considers projects that are closely entangled with both the practices and imaginaries of the spaces and experiments of positivistic science.

Revisited repeatedly throughout the chapters was the sense that, taking leave from art history and theory, from histories of geography, and from existing work on interdisciplinarity, new modes of knowledge production, consumption, and circulation are not only thoroughly intertwined but also require new modes of query and critique. In other words, to embrace the coming together of geographical research and creative practices might just in some instances reaffirm aspects of business as usual for those involved. But it might also require us to remain open to considering anew the methods and approaches through which we produce research, the means by which we assume it circulates but also the modes and terms through which we consume it and critique it. As the chapters on the page and the thesis sought to explore in particular, these are questions with implications for some of our most 'standard' academic forms: the journal article and the doctoral thesis. Forcing us to confront what we could perhaps consider to be the linguistic imperialism of our cultures of geographical research (despite the acknowledgement of geography's long history as a visual discipline) offers exciting possibilities to revisit accepted practices, to ask where we might shift these, or where they might productively stay the same. After all, as many of the practice-based PhD students I spoke to made clear, they valued the written element of their thesis, and the communicational and creative possibilities it offered. Not only did it enable their research to reach those geographers who, for very good reason, might lack a facility to appreciate practice-based output (a poem, a dance, an installation) as research, but also could function as a generative constraint, in-fact helping to advance the creative practice itself, rather than being antithetical to it.

The ideas of research, geography, and creativity that emerge are then multiple. What they direct us towards is an approach that, rather than enact fixed protocols and practices of research methods or of review, asks us to remain open to what these might be each time. How is it that we might ask of each piece of research what forms will most help circulate this to its audience, maybe that is a paper, or maybe it is something else? What is the most appropriate way to review this piece of work, maybe it is to seek an account of method, a particular approach to literature or the name-checking of theorists or practitioners, or maybe it is not. For many of us, research is perhaps already being practiced like this, but these approaches remind us of the possibilities of doing so, but also of the challenges too, and why sometimes strategic disciplinarity might be a useful thing.

The volume's third trajectory concerned how across the chapters (in some cases more than others) the understandings of research, creativity, and geography that emerged were closely tied to the ideas of the site and our relations to it that evolved. In the case, for example, of the field and the studio, we could clearly see these crucial, iconic sites of geography and art, respectively, were put into question by the projects being explored. This is, as Beth Greenhough, reflecting on the laboratory as a site for scientific knowledge production makes clear, not just to recognise the co-constitutive way in which site and knowledge emerge together but is also to appreciate how scientific practices, and in this case a range of creative practices also 'serve to question pre-existing understandings of space and the social relations sustained within them.'[5] This exploration of key sites like the field and the studio had the dubious benefit of emerging in the context of long-standing debates about these locations as sites not only where forms of knowledge making are enacted, but also as barometers for ongoing questions of what art is and what geography is. Other sites were perhaps free of this baggage but had their own challenges. One of the ongoing concerns within these discussions was that the understandings and imaginations of these sites that were emerging from the midst of the intersections of geographical research and creative practice, were not too simply set up against 'straw-man' understandings of previous relations. The chapter on laboratory, and especially the discussion of the injunction to experiment, usefully frames the need to explode reductive imaginations that tend to circulate and instead attend to the situated specificities of spaces and practices.

I framed this volume as an experiment with the GeoHumanities. As I noted, I use the word 'experiment' tentatively, especially given the concerns that have been expressed over the over-usage of such a term and the need to more carefully parse its nomadic form. I felt like the idea of the geohumanities at work here negotiated three areas of risk. First, what exactly is it, and what does it mean to conduct a geohumanities experiment? The logics of this experiment concerned less the application of something like a geohumanities protocol and more a concern to approach each site through a diversity range of methodological approaches as well as bodies of theory and ideas drawn from across the geohumanities. In a sense then the geohumanities was perhaps a disposition towards the world. A sort of way to approach knowledge that is open to drawing what is needed from various practices, approaches, and bodies of work. What working at the intersection of geographic research and creative practices perhaps means, is an orientation towards research that foregrounds the open question, that situates something called research as an open-ended question posed not only to that which is studied but also to all stages of the research process itself. The result might be to enact something already known and familiar, or it might be to ask anew what is research here, what sorts of methods are needed, what can I produce and what might it do.

Second, does the form of geohumanities experiment that this text has taken, risk, as with much work produced across disciplines, just annoying everyone? What is too rich and diverse for some, might be a delight for others. What is one reader's intellectual dilatantism might be another's generative alignment or intellectual lure? What, in short, are the terms of success of such a form? What are its modes of critique? What do we expect our interdisciplinary texts to do? In this, I can only hope that readers meet the text where it is at, and in the terms of what it was trying to do, rather than seek to judge it by more accepted forms of art history or as a form of contemporary history of geography, for in both these terms it will surely fail. In one day three departmental corridor conversations (where I am sure much of the work of academics take place these days) illustrate quandaries I seem to increasingly find myself in, wondering how to balance excitement at someone working across disciplines and practices, with critical reflection on the terms upon which this is done. I often end up feeling unfair, churlish, and grumpy. One chat concerned a CV, what do I expect the CV of someone who produces practice in a geographical context to look like? Their practice CV is not strong, and their geographical situation of their work feels weak in terms of geographical concepts. Is either of these critiques fair? In another case, PhD applications from arts and humanities disciplines were using geographical literature and ideas dating from the 1970s, for some doing the ranking, this was being praised as evidence of interdisciplinarity, for others this was seen as hopelessly dated. In a third case, PhD applications from geography were being challenged as to whether the candidates proposing exhibitions had enough training to make these work, how much was enough training? No one it seemed could win. I was left concerned that maybe we risk snuffing out this kind of research before it even gets going if we police its terms too early and too firmly. Yet I was also left discomforted by what felt like a lack of respect for disciplinary and for practice-based expertise.

My third area of risk as regards this GeoHumanities experiment was a nagging sense that the GeoHumanities might just be a 'lipstick on a pig' moment, or just another abracadabra word. In other words, is this just a new name, a linguistic flourish that sounds newly fashionable, but that really changes little of our work in practice. What, in short, did claiming this as a geohumanities experiment really let me do, that say, cultural geography did not already enable? My hunch is that perhaps it helps me think about how geohumanities is more than a body of work, or even an approach to interdisciplinarity, but is rather a reminder or perhaps even an injunction to keep questioning. What I mean by this, is that perhaps the force of the geohumanities is about staying alive to the need to keep questioning the terms upon which the compound it signals – of geography and the arts and humanities – does and is 'research.'

Denis Cosgrove, in his posthumously published prologue for the edited collection *Envisioning Landscapes, Making Worlds: Geography and the Humanities*, suggests that despite the success and ground gained by the spatial turn and by New Cultural Geography, there is still more work to be done at the intersection of geography and the humanities (and by extension I would add the arts).[6] He observes, 'the individualistic, reflective and pedagogical concerns of the Humanities remain distinct from the collective, interventionist and scientific research concerns of Social Science.'[7] He notes

> this divergence is reflected within Geography: it is apparent in the choice of materials for study, the statement of scholarly goals, the framing of

argument and the use of language more than in explicit choices of theory and method. And it remains an important part of the richness of our discipline.[8]

I don't disagree; these differences, as well as those between Humanities, Social Science, and Science, too are vital. Yet I do also wonder at the distinctions here, whether in fact from the midst of projects at the intersection of creative practices and geographic research comes a putting into question not of the differences themselves but of such a clear set of distinctions that are so easily sorted by ideas of Science, Social Science, and Humanities. With such questions, comes a performative probing of how we might do research differently, what that might make possible, what it might also of course put at risk and what we might want to ensure we value and hold dear.

I also wonder too whether we might need to not overlook the power of the cosmetic, the performative flourish of the name. We should not underestimate the energies that can gather under and around something new, the force of currents that get channelled together, at the right time in the right place. I think there is something in the way that the compound word geohumanities puts into question relations between geography and the humanities and arts, that perhaps cultural geography does not do so clearly in its naming. In other words, is it easier for those from the arts and humanities to get behind something called the geohumanities than something called cultural geography? I am not sure, but if so, and if the performative flourish opens the way for some work that requires we question the terms upon which we do research, how we value it and what it does, that has surely to be a good thing?

Notes

1 See for example Livingstone, *Putting Science in its Place.*
2 Manning and Massumi, *Thought in the Act,* p. 88.
3 Massumi, in Manning and Massumi, *Thought in the Act,* p. 188.
4 Buren, "The Function of the Studio."
5 Greenhough, "Tales of an Island-Laboratory,' p. 225.
6 Cosgrove, "Prologue," p. xxiv.
7 Ibid.
8 Ibid.

BIBLIOGRAPHY

Agnew, J. and Livingston, D. *Handbook of Geographical Knowledge*. London: Sage. 2011.

Ahmed, S. *On Being Included*. Durham, NC: Duke University Press. 2012.

Aitken, S.C. "A phenomenology of caving." *The Canadian Caver* 18 (2) (1986): 26–29.

van den Akker, P. "'Out of disegno invention is born' – drawing a convincing figure in renaissance Italian art." *Argumentation* 7 (1) (1993): 45–66.

Alfrey, N.J., Daniels. S., and Postle, M. *Art of the Garden: The Garden in British Art, 1800 to the Present Day*. London: Tate Publishing. 2004.

Arends, B. and Lebas, C. *Field Studies: Walking through Landscapes and Archives*. Amsterdam: FwBooks. 2018.

Askins, K. and Blazek, M. "Feeling our way: Academia, emotions and a politics of care." *Social and Cultural Geography* 18 (8) (2017): 1086–1105.

Askins, K. and Pain, R. "Contact zones: Participation, materiality and the messiness of interaction." *Environment and Planning D: Society and Space* 29 (5) (2011): 803–821.

Back, L. "Live sociology: Social research and its futures." *The Sociological Review* 60 (1) (2012): 18–39.

Baerwald, T.J. "Prospects for geography as an interdisciplinary discipline." *Annals of the Association of American Geographers* 100 (2010): 493–501.

Bagelman, J.J. and Bagelman, C. "Zines: Crafting change and repurposing the neoliberal university." *ACME: An International Journal for Critical Geographies* 15 (2) (2016): 365–392.

Bagelman, J.J., Nunez-Silva, M.A., and Bagelman, C. "Cookbooks: A tool for engaged research." *GeoHumanities* 3 (2) (2017): 371–395.

Bain, A. "Female artistic identity in place: The studio." *Social and Cultural Geography* 5 (2) (2004): 179–193.

Balm, R. "Expeditionary art: An appraisal." *Geographical Review* 90 (4) (2000): 585–602.

Barnett, C. "Cultural twists and turns." *Environment and Planning D: Society and Space* 16 (1998): 631–634.

Barnett, C. "Disseminating Africa: Burdens of representation and the African Writers Series." *New Formations* 57 (Winter) (2006): 74–94.

Barrett, E. and Bolt, B. *Practice as Research: Approaches to Creative Arts Enquiry*. London: IB Tauris. 2010.

Barry, A. and Born, G. (eds). *Interdisciplinarity: Reconfiguration of the Social and Natural Sciences*. London: Routledge. 2013

Bastashevski, M. "The perfect con." *E-flux* 75 (2016), unpaginated.

Bauer, L. "From bottega to studio." *Renaissance Studies* 22 (5) (2008): 642–649.

Baxter, R. "Review of modern futures." *Cultural Geographies* 24 (4) (2016): 654–656.

Beer, G. *Darwin's Plots. Evolutionary Narratives in Darwin, George Eliot and Nineteenth Century Fiction*. Cambridge: Cambridge University Press. 2009.

Ben Hayoun, N. *Homo Faber and Animal Laborans Met in Mission Control to Dream of Space: The Design of Experiences at NASA*, PhD RHUL. 2017.

Bennett, D., Blom, D., and Wright, D. Artist academics: Performing the Australian research agenda. *International Journal of Education and the Arts* 10 (17) (2009): 1–15.

Bennett, T. "The exhibitionary complex." *New Formations* 4 (1988): 73–92.

Berger, M., Chave, A.C., Kreutzer, M., Norden, L., Storr, R., and Hesse, E. *Eva Hesse: A Retrospective*. New Haven, CT: Yale University Art Gallery. 1992.

Berthoin A. "Artistic intervention residencies and their intermediaries: A comparative analysis." *Organizational Aesthetics* 1 (1) (2012): 44–67.

Berthoin, A.A. "The studio in the firm: A study of four artistic intervention residencies." In *Studio Studies: Operations, Topologies and Displacements*. Edited by Farías, I. and Wilkie, A. 175–190. London: Routledge. 2015.

Biggs, I. "The spaces of deep mapping: A partial account." *Journal of Arts and Communities* 2 (1) (2011): 5–25.

Bishop, C. *Installation Art*. London: Tate Publishing. 2005.

Bishop, C. "The social turn: Collaboration and its discontents." *Artforum* 44 (2006a): 178–183.

Bishop, C. *Participation (Documents of Contemporary Art)*. Cambridge, MA: The MIT Press. 2006b.

Bishop, C. *Artificial Hells: Participatory Art and the Politics of Spectatorship*. London: Verso. 2012.

Bishop, C. *Review Venice Biennale*, Art Forum, September 2015. https://www.artforum.com/print/201507/claire-bishop-5449 (accessed 18/12/2019).

Bishop, C. "Zones of indistinguishability: Collective actions group and participatory art." *E-flux* 29, (November 2011), unpaginated.

Bishop, C. "Information overload: Research based art and the politics of spectatorship." 2019b. Published online at http://kunsthallewien.at/#/blog/2019/01/notes-claire-bishops-lecture (accessed 18/12/2019).

Blackmore, P. and Kandiko-Howson, C. "Motivation in academic life: A prestige economy." *Research in Post-Compulsory Education* 16 (2011): 399–411.

Blok, A., Farias, I., and Roberts, C. *The Routledge Companion to Actor-Network Theory*. London: Routledge. 2020.

Bois, Y-Al., Feher, M., Foster, H., and Weizman, E. "On forensic architecture: A conversation with Eyal Weizman." *October* 156 (2016):116–140.

Bond, D.W. "Enlightenment geography in the study: A.F. Busching, J.D. Michaelis and the place of geographical knowledge in the Royal Danish Expedition to Arabia, 1761–1767." *Journal of Historical Geography* 51 (2016): 64–75.

Bond, S., DeSilvey, C., and Ryan, J.R. *Visible Mending: Everyday Repairs in the South West*. Exeter: Uniform Books. 2013.

Bourriaud, N. *Relational Art*. Paris: Les Presse du Reel. 1998.

Boyd, C. *Non-representational Geographies of Therapeutic Art Making.* London: Springer. 2017.

Boyd, C. and Edwardes, C. *Non-representational Theory and the Creative Arts.* London: Palgrave. 2019.

Brace, C. and Johns-Putra, A. "Recovering inspiration in the spaces of creative writing." *Transactions of the Institute of British Geographers.* 35 (3) (2010): 399–413.

Bracken, L. and Mawdsley, E. "Muddy glee: Rounding out the picture of women and physical geography fieldwork." *Area* 36 (3) (2004): 280–286.

Bracken, L. and Oughton, E.A. "'What do you mean?' The importance of language in developing interdisciplinary research." *Transactions of the Institute of British Geographers* 31 (2006): 371–382.

Bradberg, D. *Being There; The Necessity of Fieldwork.* Washington, DC: Smithsonian University Press. 1998.

Brice, S. "Situating skill: Contemporay observational drawing as a spatial method in geographical research." *Cultural Geographies* 25 (1) (2017): 135–158.

Brickell, K. "Participatory video drama research in transitional Vietnam: Post-production narratives on marriage, parenting and domestic evils." *Gender, Place and Culture* 22 (4) (2015): 510–525.

Brickell, K., Cristofoletti, T., Chann, S., Natarajan N., and Parsons, L. "Exhibition - blood bricks: Untold stories of modern slavery and climate change from Cambodia." *Environment and Planning D Online Magazine.* 2019. https://www.societyandspace.org/ articles/blood-bricks-untold-stories-of-modern-slavery-and-climate-change-from-cambodia (accessed 18/12/2019).

Buchanan, B., Chrelew, M., and Bastian, M. "Issue 4: Field philosophy and other experiments." *Parallax* 24 (4) (2018): 383–391.

Buller, H. "The lively process of interdisciplinarity." *Area* 41 (2008): 395–403.

Buren, D. "The function of the studio." In *The Studio Reader: On the Space of Artists.* Edited by Jacob, M.J. and Grabner, M. 156–162. Chicago, IL: University of Chicago Press. 1971.

Burke, M. "Knitting the atmosphere: Creative entanglements with climate change." In *Geographies of Making, Craft and Creativity.* Edited by Price, L. and Hawkins, H. 158–173. London: Routledge. 2018.

Butt, D. *Artistic Research in the Future Academy.* Chicago, IL: Intellect/ University of Chicago Press. 2019.

Callard, F. and Fitzgerald, D. *Rethinking Interdisciplinarity across the Social Sciences and Neurosciences.* Basingstoke: Palgrave Macmillan. 2015.

Candlin, F. "A proper anxiety: Practice-based PhDs and academic unease." *Working Papers in Art and Design* 1 (2000), unpaginated.

Candlin, F. "A dual inheritance: The politics of educational reform and PhDs in art and design." *The International Journal of Art and Design Education* 20 (3) (2008): 302–310.

Candy, L. "Differences between practice-based and practice-led research." Creativity and cognition studios report. University of Technology, Sydney. 2006. http://www.creativityandcognition.com/content/view/124/131 (accessed 18/12/2019).

Cant, S. "The tug of danger with the magnetism of mystery Descents into 'the comprehensive poetic-sensuous appeal of caves'." *Tourist Studies* 3 (1) (2003): 67–81.

Caprotti, F. and Cowley, R. "Interrogating urban experiments." *Urban Geography* 28 (9) (2017): 1441–1450.

Castree, N., Fuller, D., and Lambert, D. "Boundary crossings: Geography without borders." *Transactions of the Institute of British Geographers* 31 (2007): 129–132.

Cazeaux, C. *Art, Research, Philosophy.* London: Routledge. 2016.

Chapman, H.P. "The imagined studios of Rembrandt and Vermeer." In *Inventions of the Studio, Renaissance to Romanticism.* Edited by Cole, M. and Pardo, M. 108–146. Chapel Hill: University of North Carolina Press. 2005.

Clifford, J. and Marcus, G. (eds). *Writing Cultures: The Politics and Poetics of Ethnography.* Berkley: University of California Press. 1986.

Clifford, N. "Physical geography: The naughty world revisited." *Transactions of the Institute of British Geographers* 26 (2001): 387–389.

Cochrane, A. and Russell, I.A. (eds). *Art and Archaeology: Collaborations, Conversations, Criticisms.* One World Archaeology Series, Volume 11. New York: Springer-Kluwer. 2014.

Coles, A. *The Transdisciplinary Studio.* Berlin: Sternberg Press. 2012.

Cole, M. and Pardo, M. "Origins of the studio." In *Inventions of the Studio, Renaissance to Romanticism.* Edited by Cole, M. and Pardo, M. 1–35. Chapel Hill: University of Northern Carolina Press. 2005a.

Cole, M. and Pardo, M. (eds). *Inventions of the Studio, Renaissance to Romanticism.* Chapel Hill: University of Northern Carolina Press. 2005b.

Condee, W. "The interdisciplinary turn in the arts and humanities." *Issues in Interdisciplinary Studies* 34 (2016): 12–29.

Cook, I. et al. "'Afters': 26 Authors and a 'workshop imagination geared to writing,'" *Cultural Geographies* 21 (1) (2013): 135–140.

Cook, I. et al., "A new vocabulary for cultural-economic geography?" *Dialogues in Human Geography* 9 (1) (2019): 83–87.

Corcoran, K., Delfos, C., and Maxwell, J. (eds). *ArtFutures: Working with Contradictions in Higher Arts Education.* Amsterdam: ELIA. 2014.

Cosgrove, D. "Prologue." In *Envisioning Landscapes, Making Worlds: Geography and the Humanities.* Edited by Daniels, S., DeLyser, D., Entrikin, J.N., and Richardson, D. 5–9. London: Routledge. 2012.

Crang, P. "Cultural geography, after a fashion?" *Cultural Geography* 17 (2010): 191–201.

Crang, P. and Mitchell, D. "Editorial." *Cultural Geographies* 7 (1) (2000): 1–7.

Cresswell, T. *Soil.* London: Penned in the Margins. 2013.

Cresswell, T. *Maxwell Street: Writing and Thinking Place.* Chicago, IL: University of Chicago Press. 2019.

Crouch, D. *Flirting with Space: Journeys and Creativity.* Farnham: Ashgate. 2010.

Crouch, D. "Choreo-graphic figures." *Social and Cultural Geographies* 20 (8) (2019): 1182–1184.

Cutler, A. *Nostalgia Forest.* London: Oyster Catcher Press. 2013.

Daniels, S. *Fields of Vision: Landscape Imagery and National Identity in England and the United States.* Princeton: Princeton University Press. 1993.

Daniels, S. "Review of bipolar." *Cultural Geographies* 18 (3) (2011a): 424–425.

Daniels, S. "Art studio." In *The Handbook of Geographical Knowledge.* Edited by Agnew, J. and Livingston, D. 137–148. London: Sage. 2011b.

Daniels, S., DeLyser, D., Entrikin, J.N., and Richardson, D. (eds). *Envisioning Landscapes, Making Worlds: Geography and the Humanities.* London: Routledge. 2012.

Davidts, W. and Paice, K. (eds). *The Fall of the Studio: Artists at Work.* Amsterdam: Antennae, Valiz. 2009.

Davies, G. "Where do experiments end." *Geoforum* 41 (2010): 667–670.

Dear, M., Ketchum, J., Luria, S., and Richardson, D. (eds). *GeoHumanities: Art, History and Text at the Edge of Place.* London: Routledge. 2011.

De Leeuw, S. *Geographies of a Lover.* Edmonton: NeWest Press. 2012.

De Leeuw, S. *Skeena*. Sechelt: Caitlin Press. 2015.

De Leeuw, S. and Hawkins, H. "Critical eminist geographies and creative re/turns: Poetics and practices for new disciplinary spaces." *Gender, Place and Culture* 24 (3) (2017): 303–324.

De Leeuw, S., Parkes, M.W., Morgan, V.S., Christensen, J., Lindsay, N., Mitchell-Foster, K., and Jozkow, J.R. "Going unscripted: A call to critically engage storytelling methods and methodologies in geography and the medical-health sciences." *Canadian Geographer* 61 (2) (2017): 152–164.

DeLyser, D. and Greenstein, P. "The devotions of restoration: Materiality, enthusiasm, and making three 'Indian Motocycles' like new." *Annals of the American Association of Geographers* 107 (6) (2017): 1461–1478.

DeLyser, D. and Hawkins, H. "Introduction: Writing creatively - process, practice, and product." *Cultural Geographies* 21 (1) (2014): 131–134.

Demos, T.J. "Rethinking site-specificity." *Art Journal* 62 (2) (2013): 98–100.

Derrida, J. (ed). *Of Grammatology*. Trans. Baltimore, MD: John Hopkins University Press. 1998.

Desilvey, C., Naylor, S., and Sackett, C. (eds). *Anticipatory History*. Exeter: Uniform Books. 2011.

Desilvey, C. "Mining memories: Placing the anthropocene by Gareth Hoskins and Mark Whitehead." *Cultural Geographies* 22 (2) (2015): 377–378.

Dewsbury, J.D. "Performative, non-representational, and affect-based research: Seven injunctions." In *The SAGE Handbook of Qualitative Geography*. Edited by Delyser, D., Herbert, S., Aitken, S., Crang, M., and McDowell, L. 321–334. London: SAGE, 2010.

Dewsbury, J.D. and Naylor, S. "Practising geographical knowledge: Fields, bodies and dissemination." *Area* 34 (3) (2002): 253–260.

Deutsche, R. *Evictions: Art and Spatial Politics*. Cambridge, MA: MIT Press. 1996.

Dixon, D.P. "The blade and the claw: Science, art and the creation of the lab-bourne monster." *Social and Cultural Geography* 9 (6) (2008): 671–692.

Dixon, D., Creswell, T., Bol, P., and Entrkin, N. "Editorial." *GeoHumanities* 1 (1) (2015): 1–19.

Dixon, D., Hawkins, H., and Straughan, L. "Sublimity, formalism and the place of art within geomorphology." *Progress in Physical Geography* 37 (2013): 227–247.

Doherty, C. (ed). *From Studio to Situation*. London: Black Dog Publishing. 2004.

Domosh, M. "With 'stout boots and a stout heart': Feminist methodology and historical geography." In *Thresholds in Feminist Geography*. Edited by Jones, J.-P., Nast, H., and Roberts, S. 225–240. New York: Rowman and Littlefield. 1997.

Donadelli, G., Gallanti, C., Rocca, K., and Varotto, M. "The past for the future of geography: Introducing the Padua museum of geography." 2018. Accessed https://www.musei.unipd.it/sites/musei.unipd.it/files/Donadelli%20et%20al.%20(2018b).%20The%20past%20for%20the%20future.%20Introducing%20the%20Padua%20Museum%20of%20Geography.pdf.

Donaldson, A., Ward, N., and Bradley, S. "Mess among disciplines: Interdisciplinarity in environmental research." *Environment and Planning A* 42 (2010): 1521–1536.

Driver, F. "Editorial: Fieldwork in geography." *Transactions of the Institute of British Geographers* 25 (2000): 267–268.

Driver, F. *Geography Militant: Cultures of Exploration and Empire*. London: Blackwell. 2001.

Driver, F. *The Active Life: The Explorer as Biographical Subject*. Oxford Dictionary of National Biography. 2005. http://www.oxforddnb.com/view/theme/94053 (accessed 10/07/2012).

Driver, F. "Hidden histories made visible? Reflections on a geographical exhibition." *Transactions of the Institute of British Geographers* 28 (3) (2013): 420–435.

Driver, F. and Jones, L. *The Hidden Histories of Exploration: Exhibiting Geographical Collections.* London: Royal Holloway RGS-IBG. 2009.

Driver, F. and Martins, L. *Tropical Visions in the Age of Empire.* Chicago, IL: University of Chicago Press. 2005.

Driver, F., Nash, C., Prendergast, K., and Swenson, I. *Landing: Eight Collaborative Projects between Artists + Geographers.* London: Royal Holloway. 2005.

Drucker, J. *The Century of Artists' Books.* New York: Granary Books Inc. 2004.

Duncombe, S. *Notes from the Underground. Zines and the Politics of Alternative Culture.* Portland, OR: Microcosm Publishing, 1997.

Edensor, T. "Review: Visible mending." *Cultural Geographies* 21 (4) (2014): 749–750.

Edensor, T., Leslie, D., Millington, S., and Rantisi, N. *Spaces of Vernacular Creativity: Rethinking the Cultural Economy.* London: Routledge. 2010.

El Khoury, T. "Sexist and racist people go to the theatre too." *Performing Ethos* 3 (2) (2016): 209–212.

El Khoury, T. "The audience dug the graves: Interacting with oral history and mourning in live art." Doctoral Thesis. 2018. Available at https://pure.royalholloway.ac.uk/portal/en/publications/the-audience-dug-the-graves-interacting-with-oral-history-and-mourning-in-live-art(32bba2d2-6d8b-44b6-99bd-d0c556665159).html (accessed 19/12/2019).

El Khoury, T. and Pearson, D. "Two live artists in the theatre." *Performance Research* 20 (4) (2015): 122–126.

Eliasson, O. "Welcome." In *Take Your Time Vol. 7. Open House.* Edited by Eliasson, S.O. 9–11. Cologne: Verlag der Buchhandlung Walther Konig. 2017.

Elkins, J. "The new PhD in studio art." Printed Project. Issue 4, Sculptors Society of Ireland. 2005.

Elkins, J. *Artists with PhDs: On the New Doctoral Degree in Studio Art.* New York: New Academic Publishing. 2009.

Engelmann, S. "Social spiders and hybrid webs at studio Tomás Saraceno." *Cultural Geographies* 24 (1) (2017a): 161–169.

Engelmann, S. "The cosmological aesthetics of Tomas Saraceno's atmospheric experiments." 2017b. https://ora.ox.ac.uk/objects/uuid:481978b1-eb92-4fa9-a514-2bcbbfdd7612.

Engelmann, S. "Of spiders and simulations: Art-machines at Studio Tomás Saraceno." *Cultural Geographies* 26 (3) (2019): 305–322.

Enigbokan, A. and Patchett, M. "Speaking with specters: Experimental geographies in practice." *Cultural Geographies* 19 (2012): 535–546.

Evans, A. *The Myth Gap: What Happens when Evidence and Arguments Aren't Enough.* Cornwall: Eden Project Books. 2017.

Feiss, E.C. "Response to Grant Kester's 'The device laid bare'." *E-flux Journal* 54 (2014) unpaginated.

Fer, B. "Objects beyond objecthood." *Oxford Art Journal* 22 (2) (1999): 27–36.

Fer, B. *Eve Hesse Studiowork.* Edinburgh: Fruitmarket Gallery. 2009.

Fer, B.A. "Eva Hesse: Painting drawing sculpture." In *Eva Hesse: Spectres and Studiowork.* Edited by Hesse, A. and McKinnon, L. 10–19. Seoul: Kukje Gallery. 2012.

Ferguson, B.W., Greenberg, R., and Naire, S. "Mapping international exhibitions." In *Thinking about Exhibitions.* Edited by Greenberg, R. Ferguson, B., and Nairne, S. London and New York: Routledge. 1996.

Flintham, M. "Visualising the invisible: Artistic methods towards military airspaces." In *The Routledge Companion to Military Research Methods*. Edited by Williams, A.J., Jenkings, N., Woodward, R., and Rech, M.F. 357–369. London: Routledge. 2016.

Foster H. "The artist as ethnographer?" In *The Traffic in Culture. Refiguring Art and Anthropology*. Edited by Marcus, G. and Myers, F. 302–309. Berkeley and London: University of California Press. 1995.

Foster, H. "Archival impulse." *October* 110 (2004): 3–22.

Foster, K. and Lorimer, H. "Some reflections on art-geography as collaboration." *Cultural Geographies* 14 (2007): 425–432.

Fraser, T. "The studio: History, myth and legacy. On the changing nature of the space used by artists and the work being done there." On, http://www.terriefraser.com/the_studio.pdf (accessed 19/12/2019).

Gallagher, M. "Field recording and the sounding of spaces." *Environment and Planning D: Society and Space* 33 (3) (2015): 560–576.

Gallagher, M. and Prior, J. "Sonic geographies: Exploring phonographic methods." *Progress in Human Geography* 38 (2) (2014): 267–284.

Gallagher, M., Kanngieser, A., and Prior, J. "Listening geographies: Landscape, affect and geotechnologies." *Progress in Human Geography* 41 (5) (2018): 618–637.

Galison, P. "Three laboratories." *Social Research* 64 (3) (1997): 1127–1155.

Galison, P. "Buildings and the subjects of science." In *The Architecture of Science*. Edited by Galison, P. and Thompson, E. 1–29. Boston, MA: MIT Press. 1999.

Galison, P. and Jones, C. "Factory, laboratory, Studio. Dispersing sites of production." In *The Architecture of Science*. Edited by Galison, P. and Thompson, E. 497–540. Boston, MA: MIT Press. 1999.

Galison, P. and Thompson, E. *The Architecture of Science*. Boston, MA: MIT Press. 1999.

Garrett, B. "Videographic geographies: Using digital video for geographic research." *Progress in Human Geography* 35 (2011): 521–541.

Gat, O. "Could reading be looking?" *E-flux* 72 (2016): 1–9.

Genosko, G. *Felix Guattari: An Aberrant Introduction*. London: Bloomsbury. 2012.

Gertner, J. *The Idea Factory: Bell Labs and the Great Age of American Innovation*. New York: Penguin.

Gibson, C. and Klocker, N. "Academic publishing as 'creative' industry, and recent discourses of 'creative econoimes': Some critical reflections." *Area* 36 (4) (2004): 423–434.

Gielen, P. "Community art: A neo-liberal solution for the deconstruction of welfare state? Respublika! Experiments in the performance of participation and democracy. Cyprus, NeMe. 67–81. 2019.

Gkartzios, M. and Crawshaw, J. "Researching rural housing: With an artist in residence." *Sociologia Ruralis* 59 (4) (2019): 589–611.

Glinkowski, P. and Bamford, A. *Insight and Exchange: An Evaluation of the Wellcome Trust's Sciart Programme*. London: Wellcome Trust. 2009. www.wellcome.ac.uk/sciartevaluation (accessed 01/11/2009).

Gorman-Murray, A. and Brickell, C. "Over the ditch: Queer mobilities at the nexus of art, geography and history." *ACME* 16 (3) (2017): 576–604.

Gotlieb, M. *Creation and Death in the Romantic Studio. Inventions of the studio: Renaissance to Romanticism*. Chapel Hill and London: University of North Carolina Press. 2005.

Greenberg, C. "Modernist painting." In *Clement Greenberg: The Collected Essays and Criticism, Volume 4: Modernism with a Vengeance, 1957–1969*. Edited by O'Brian, J. 85–93. Chicago, IL: University of Chicago Press. 1993.

Greenhough, B. "Tales of an Island-laboratory: defining the field in geography and science studies." *Transactions of the Institution of British Geographers* 31 (2006): 224–237.

Greppi, C. "'On the spot' travelling artists and the iconographic inventory of the world, 1769–1859." In *Tropical Visions in the Age of Empire*. Edited by Driver, F. and Martins, L. 23–42. Chicago, IL: University of Chicago Press. 2005.

Gross, M. *Ignorance and Surprise: Science, Society, and Ecological Design*. Cambridge, MA: MIT Press. 2010.

Grzelec, A. and Prata, T. "Artists in organisations: Mapping of European producers of artistic interventions in organisations." Creative Clash report. Gothenburg, Sweden: TILLT Europe. 2013. http://www.creativeclash.eu/wpcontent/uploads/2013/03/Creative_Clash_Mapping_2013_GrzelecPrata3.pdf (accessed 19/12/2019).

Hammarfelt, B. "Using altmetrics for assessing research impact in the humanities." *Scientometrics* 101 (2) (2014): 1419–1430.

Hannula, M., Kaila, J., Palmer, R., and Sarje, K. (eds). *Artists as Researchers: A New Paradigm for Art Education in Europe*. Helsinki: Academy of Fine Arts, Helsinki. 2013.

Haraway, D. "Situated knowledges: The science question in feminism and the privilege of partial perspective." In *Human Geography: An Essential Anthology*. 108–128. Edited by Agnew, J., Livingstone, D., and Rogers, A. Oxford: Blackwell. 1996.

Harel-Shalev, A., Huss, E., Daphna-Tekoah, S., and Cwikel, J. "Drawing (on) women's military experiences and narratives – Israeli women soldiers' challenges in the military environment." *Gender, Place & Culture* 24 (4) (2017): 499–514.

Harman, S. *Seeing Politics: Film, Visual Method, and International Relations*. Kingston: McGill Queens University Press. 2019.

Harris, C. *Art and Innovation: The Xerox PARC Artist-in-Residence Program*. Cambridge, MA: MIT Press. 1999.

Harvie, J. *Fair Play-Art, Performance and Neoliberalism*. London: Palgrave. 2013.

Haseman, B. and Mafe, D. "Acquiring know-how: Research training for practice-led researchers." In *Practice-Led Research, Research-Led Practice in the Creative Arts*. Edited by Smith, H. and Dean, R.T. 211–228. Edinburgh: Edinburgh University Press. 2009.

Hawkins, H. "The argument of the eye'? The cultural geographies of installation art." *Cultural Geographies* 17 (3) (2010): 321–340.

Hawkins, H. "Dialogues and doings: Geography and art-landscape, critical spatialities and participation." *Geography Compass* 5 (2011): 464–478.

Hawkins, H. *For Creative Geographies: Geography, Visual Arts and the Making of Worlds*. New York: Routledge. 2014.

Hawkins, H. *Creativity: Live, Work, Create*. London: Routledge. 2016.

Hawkins, H. "Creative geographic methods: Knowing, representing, intervening. On composing place and page." *Cultural Geographies* 22 (2) (2015): 247–268.

Hawkins, H. "Geography's creative (re)turn: Toward a critical framework." *Progress in Human Geography* 43 (6) (2018): 963–984.

Hawkins, H. "Doing gender and the GeoHumanities - Celebrations and intoxications." *Gender, Place and Culture* 26 (11) (2019): 1503–1518.

Hawkins, H. "Underground imaginations, environmental crisis and subterranean cultural geographies." *Cultural Geographies* 27 (1) (2020): 3–22.

Hawkins, H. and Catlow, R. "Shaping subjects, connecting communities, imagining futures? Critically investigating play your place." In *Public Art Encounters*. Edited by Zebracki, M. and J. Palmer, J. 91–107. London and New York: Routledge. 2018.

Hawkins, H., Marston, S.A., Ingram, M., and Straughan, E. "The art of socioecological transformation." *Annals of the Association of American Geographers* 105 (2) (2015): 331–341.

Hayes, E. "Slidescapes: Three royal geographical society lantern lectures by Vaughan Cornish." *Early Popular Visual Cultures* 17 (1) (2019): 71–88.

Hediger, I. and Scott, J. *Swiss Artists in Labs: Recomposing Art and Science.* Berlin: De Gruyter. 2016.

van Heeswijk, J. 2012. "The artist will have to choose whom to serve." In *Social Housing - Housing the Social: Art, Property and Spatial Justice.* Edited by Phillips, A. and Erdemci, F. 78. Berlin: Sternberg Press/SKOR.

Hemer, S R. "Coffee as a supervisory technique; power, formality and the persona in supervision." 2009. Available on line from https://www.adelaide.edu.au/carst/docs/hemer-2009-supercino.pdf (accessed 27/08/2016).

Henderson, E.F. "Academic conferences: Representative and resistant sites for higher education research." *Higher Education Research and Development* 34 (5) (2015): 914–925.

Hercombe, S. *What the Hell Do We Want an Artist Here For?* London: Calouste Gulbenkian Foundation. 1986.

Hetherington, P. (ed). *Issues in Art and Education: Artists in the 1990s, Their Education and Values.* London: Tate Publishing. 1994.

Higgin, M. "What do we do when we draw?" *Tracey*, www.lboro.ac.uk/microsites/sota/tracey (accessed 19/12/2019).

Higins, K. "Inspiration and exchange: Artist residencies in Oceania." PhD Thesis, University of Auckland. 2012. https://researchspace.auckland.ac.nz/handle/2292/17387 (accessed 19/12/2019).

Hinchliffe, S. "Indeterminacy in-decisions – Science, policy and politics in the BSE (bovine spongiform encephalopathy) crisis." *Transactions of the Institute of British Geographers* 26 (2001): 182–204.

Hinchliffe, S., Kearnes, M.B., Degan, M., and Whatmore, S. "Urban wild things: A cosmopolitical experiment." *Environment and Planning D: Society and Space.* 23 (5) (2005): 643–658.

Hockey, J. "Establishing boundaries: Problems and solutions in managing PhD supervisor's role." *Cambridge Journal of Education* 24 (2) (1994): 289–305.

Hockey, J. "United Kingdom art and design practice-based PhDs: Evidence from students and their supervisors." *Studies in Art Education* 48 (2) (2007): 155–171.

Holert, T. "Evaluating knowledge= evaluating art?" *Texte Zur Kunst.* 2018. https://www.textezurkunst.de/articles/evaluating-knowledge-evaluating-art/.

Holmes, D. and Marcus, G.E. "Cultures of expertise and the management of globalization: Toward the refunctioning of ethnography." In *Global Assemblages: Technology, Politics and Ethics as Anthropological Problems.* Edited by Ong, A. and Collier, S.J. 235–252. London: Blackwell. 2005.

Holmes, D. and Marcus, G.E. "Fast capitalism: Para-ethnography and the rise of the symbolic analyst." In *Frontiers of Capital: Ethnographic Reflections on the New Economy.* Edited by Melissa Fisher, M. and Downey, G. 33–57. Durham, NC: Duke University Press. 2006.

Hoskins, G. and Whitehead, M. *Mining Memories: Placing the Anthropocene.* Aberystwyth: Prifysgol Aberystwyth, Aberystwyth University. 2013.

Hubbard, P. "Review of modern futures." *Journal of Historical Geography* 57 (2017): 121–122.

Hudek, A. "The incidental person." 2010. http://www.apexart.org/exhibitions/hudek.php (accessed 19/12/2019).

Hughes, R. "The poetics of practice-based research writing." *The Journal of Architecture* 11 (3) (2006): 283–301.

Ingold, T. "Anthropology is not ethnography." *Proceedings of the British Academy* 154 (2008a): 69–92.

Ingold, T. "Bindings against boundaries: Entanglements of life in an open world." *Environment and Planning A: Economy and Space* 40 (8) (2008b): 1796–1810.

Ingold, T. *Being Alive: Essays on Movement, Knowledge and Description.* Abingdon: Routledge. 2011.

Ingold, T. *Making: Anthropology, Archaeology, Art and Architecture.* London: Routledge. 2013.

Ingold, T. *Redrawing Anthropology: Materials, Movements, Lines.* London: Routledge. 2016.

Ingram, A. "Experimental geopolitics: Wafaa Bilal's domestic tension." *The Geographical Journal* 178 (2012): 123–133.

Ingram, M. "Washing urban water: Diplomacy in environmental art in the Bronx, New York City." *Gender, Place & Culture: A Journal of Feminist Geography* 21 (1) (2013): 105–122.

Jackson, S. *Social Works: Performing Arts, Supporting Publics.* London: Routledge. 2011.

Jacob, M.J. and Grabner, M. (eds). *The Studio Reader: On the Space of Artists.* Chicago and London: University of Chicago Press. 2012.

Jellis, T. "Reclaiming experiment: geographies of experiment and experimental geographies." 2013. https://ora.ox.ac.uk/objects/uuid:39de7269-7ddf-4aaa-a4a1-ae6ad9ed17bb.

Jellis, T. "Spatial experiments: Art, geography, pedagogy." *Cultural Geographies* 22 (2) (2015): 369–374.

Jones, C.A. *Machine in the Studio: Constructing the Postwar American Artist.* Chicago, IL: University of Chicago Press. 1997.

Jones, C.A. "The server/user mode: The art of Olafur Eliasson." *Artforum International* 46 (2) (2007): non-paginated online.

Jones, C.A. "Event Horizon: Olafur Eliasson's raumexperimente." In *Olafur Eliasson: Contact.* Edited by Eliasson, S.O. 132–137. Berlin: Flammarion. 2014.

Katz, C. "Playing the field: Questions of fieldwork in geography." *Professional Geography* 46 (1994): 67–72.

Keighren, I.M. *Bringing Geography to Book: Ellen Semple and the Reception of Geographical Knowledge.* London: IB Tauris. 2010.

Keighren, I.M. "Geographies of the book: Review and prospect." *Geography Compass* 7 (11) (2013): 745–58.

Keighren, I.M., Withers, C.W.J., and Bell, B. *Travels into Print: Exploration, Writing, and Publishing with John Murray, 1773–1859.* Chicago, IL: University of Chicago Press. 2015.

Kester, G. "The device laid bare: On some limitations in current art criticism." *E-flux Journal* 50 (2013). http://www.e-flux.com/journal/the-device-laid-bare-on-some-limitations-in-current-ar t-criticism/ (accessed 19/12/2019).

Kindon, S., Pain, R., and Kesby, M. *Participatory Action Research Approaches and Methods: Connecting People, Participation and Place.* London: Routledge. 2007.

Kinpaisby-Hill, C. "Participatory praxis and social justice: Towards more fully social geographies." In *A Companion to Social Geography.* Edited by Del Casino, V., Thomas, M., Cloke, P., and Panelli, R. 214–234. London: Blackwell. 2011.

Kirsch, S. "Laboratory/observatory." In *Sage Handbook of Geographical Knowledge.* Edited by Agnew, J. and Livingstone, S. 76–87. London: Sage. 2011.

Kitchin, R. "Engaging publics: Writing as praxis." *Cultural Geographies* 21 (1) (2013): 153–157.

Klein, J. "What is artistic research? Reflections." *JAR: Journal for Artistic Research* 2017. https://www.jar-online.net/what-artistic-research (accessed 19/12/2019).

Klonk, C. *New Laboratories: Historical and Critical Perspectives on Contemporary Developments.* Berlin: De Gruyter. 2016.

Knorr-Cetina, K.D. *Epistemic Cultures: How the Sciences Make Knowledge*. Harvard: London University Press. 1999.

Kohler, R.E. *Landscapes and Labscapes. Exploring the Lab-Field Border in Biology*. Chicago, IL: University of Chicago Press. 2002.

Kohler, R.E. "Lab history: Reflections." *Isis* 99 (4) (2008): 761–768.

Konrad, M. (ed). *Collaborators Collaborating: Counterparts in Anthropological Knowledge and International Research Relations*. New York and Oxford: Berghahan Books. 2012.

Kroll, J. "Creative writing as research and the dilemma of accreditation: How do we prove the value of what we do?" *Text Journal* April (2004), unpaginated.

Krupar, S. *Hot Spotter's Report: Military Fables of Toxic Waste*. Minneapolis and London: University of Minnesota Press. 2013.

Krysa, J. "Exhibitionary practices at the intersection of academic research and public display." In *The Futures of Artistic Research*. Edited by Seppä, A., Slager, H., and Kaila, J. 63–72. Helsinki: University of the Arts Helsinki. 2017.

Kullman, K. "Geographies of experiment/experimental geographies. A rough guide." *Geography Compass* 7 (12) (2013): 879–894.

Kwon, M. *One Place after Another: Site Specific Art and Locational Identity*. Cambridge, MA: The MIT Press. 2002.

Lacey, S. *Mapping the Terrain. New Genre Public Art*. Seattle, WA: Bay Press. 1995.

Lane, S.N. "Constructive comments on D Massey 'space-time,' 'science' and the relationship between physical and human geography." *Transactions of the Institute of British Geographers* 26 (2001): 243–256.

Last, A. "Experimental geographies." *Geography Compass* 6 (2014): 706–724.

Latham, J. "Artist: John Latham placement: Scottish Office (Edinburgh)." *Studio International* March-April (1976): 169–170.

Latham, A. and Conradson, D. "The possibilities of performance." *Environment and Planning A: Ecnomy and Space* 35 (11) (2003): 1901–1906.

Latham, A. and McCormack, D.P. "Thinking with images in non-representational cities: Vignettes from Berlin." *Area* 41 (3) (2009): 252–262.

Latour, B. *Reset Modernity! Field Book*. Cambridge, MA: MIT Press. 2016. Also available to download from: http://www.bruno-latour.fr/sites/default/files/downloads/RESET-MODERNITY-GB.pdf (accessed 19/12/2019).

Leaver-Yap, I. "Eva Hesse: Present tense." *MAP* 19 (Autumn) (2009). https://mapmagazine.co.uk/eva-hesse-present-tense (accessed 19/12/2019).

Lebas, C. *Field Studies: Walking through Landscapes and Archives*. Berlin: FW-Books 2018.

Lederman, R. "The perils of working at home: IRB 'mission creep's context and content for an ethnography of disciplinary knowledge." *American Ethnologist* 33 (4) (2006): 482–491.

Lee, B. Fillis, I. and Lehman, K. "Art, science and organisational interactions: Exploring the value of artistic residencies on campus." *Journal of Business Research* 85 (2018): 444–451.

Lesage, D. "Who's afraid of artistic research? On measuring artistic research output." *Art & Research: A Journal of Ideas, Contexts and Methods* 2 (2) (2009): 1–10.

Leyshon, A. "The software slump? Digital music, the democratization of technology, and the decline of the recording studio sector within the musical economy." *Environment and Planning A* 41 (6) (2009): 1309–1331.

Liinamaa, S. "Negotiating a 'Radically Ambiguous World': Planning for the future of research at the art and design university." *International Journal of Art & Design Education* 37 (3) (2017): 426–437.

Linder, B. "Review-Maxwell street: Writing and thinking place." *Social & Cultural Geography* 21 (2) (2020): 289–291.

Lippard, L.R. *Six Years: The Dematerialisation of the Art Object from 1996 to 1972*. Berkeley: University of California Press. 1973.

Lippard, L.R. and Chandler, J. "The dematerialisation of art." *Art International* 12 (2) (1968): 31–36.

Lithgow, M. and Wall, K. "Embedded aesthetics: Artist-in-residencies as sites of discursive struggle and social innovation." *Seismopolite* 19 (2017). https://www.research gate.net/profile/Karen_Wall3/publication/321716050_Lithgow_M_and_Wall_K_2017_Embedded_Aesthetics_Artist-in-Residencies_as_Sites_of_Discursive_Struggle_and_Social_Innovation_Seismopolite_19_httpwwwseismopolite-comembedded-aesthetics-artist-in-residencie/links/5dcaee4692851c818049deba/Lithgow-M-and-Wall-K-2017-Embedded-Aesthetics-Artist-in-Residencies-as-Sites-of-Discursive-Struggle-and-Social-Innovation-Seismopolite-19-http-wwwseismopolite-com-embedded-aesthetics-artist-in-resi.pdf (accessed 19/12/2019).

Liu, A. "The power of formalism: The new historicism." *English Literary History* 56 (1989): 721–777.

Livingstone, D.N. *Putting Science in its Place. Geographies of Scientific Knowledge*. London: University of Chicago Press. 2002.

Livingstone, D.N. "Science, text and space: Thoughts on the geography of reading." *Transactions of the Institute of British Geographers* 30 (4) (2005): 391–401.

Livingstone, D.N. and Withers, C.W.J. (eds). *Geography and Enlightenment*. Chicago, IL: University of Chicago Press. 1999.

Livingstone, D.N. and Withers, C.W.J. (eds). *Geographies of Nineteenth-Century Science*. Chicago, IL: University of Chicago Press. 2011.

Longhurst, R., Ho, E., and Johnston, L. "Using 'the body' as an 'instrument of research': kimch'i and pavlova." *Area* 40 (2) (2008): 208–217.

Lorimer, H. and Foster, K. *Cross-Bills*. 2005. http://diffusion.org.uk/?p=86 (accessed 19/12/2019).

Lorimer, J. and Driessen, C. "Wild experiments at the Oostvardersplassen: Rethinking environmentalism for the anthropocene." *Transactions of the Institute of British Geographers* 39 (2) (2014): 169–181.

Lorimer, J., Hodgetts, T., Grenyer, R., Greenhough, B., McLeod, C. and Dwyer, A. "Making the microbiome public: Participatory experiments with DNA sequencing in domestic kitchens." *Transactions of the Institute of British Geographers* 44 (3) (2019): 524–541.

Lovejoy, A. "…here nor there…An artist-led initiative which celebrates the 'twinning' of two cities." *Ecumene* 7 (3) (200): 345–346.

Lovejoy A. and Hawkins, H. *Insites: An Artists' Book*. Penryn: Insites Press. 2009.

Loveless, N.S. "Practice in the flesh of theory: Art, research, and the fine arts PhD." *Canadian Journal of Communication* 3 (1) (2012): 93–108.

Macleod, K. "The functions of the written text in practice-based PhD submissions." *Working Papers in Art and Design* 1 (2000). http://www.herts.ac.uk/artdes/research/papers/wpades/vol1/macleod2.html.

Macload, K. and Chapman, N. "The absenting subject: Research notes on PhDs in fine art." *The Journal of Visual Art Practice* 13 (2) (2014): 138–149.

Macleod, K. and Holdridge, L. "The doctorate in fine art: The importance of exemplars to the research culture." *The International Journal of Art and Design Education* 23 (2) (2004): 155–168.

Maddrell, A. *Complex Locations: Women's Geographical Work in the UK 1850–1970*. London: Wiley-Blackwell. 2010.

Madge, C. "On the creative (re)turn to geography: Poetry, politics and passion." *Area* 46 (2014): 178–185.

Magee, P. "What distinguishes scholarship from art?" *New Writing: International Journal for the Practice and Theory of Creative Writing* 11 (3) (2014): 400–406.

Magrane, E. and Cokinos C. *The Sonoran Desert: A Literary Field Guide.* Tucson: University of Arizona Press. 2016.

Magrane, E., Russo, L., de Leeuw, S., and Santos-Perez, C. (eds). *Geopoetics in Practice.* London: Routledge. 2019.

Mäkelä, M. "Knowing through making: The role of the artefact in practice-led research." *Knowledge, Technology & Policy* 20 (3) (2007): 157–163.

Manathunga, C. "Supervision as mentoring: the role of power and boundary crossing." *Studies in Continuing Education* 29 (2) (2007): 207–221.

Manning, E and Massumi, B. *Thought in the Act: Passages in the Ecology of Experience.* Minneapolis: Minnesota University Press. 2014.

Marcus, G.E. "Ethnography two decades after writing culture: From the experimental the baroque." *Anthropological Quarterly* 80 (4) (2007): 1127–1145.

Marcus, G.M. "Contemporary fieldwork aesthetics in art and anthropology: Experiments in collaboration and intervention." *Visual Anthropology* 23 (4) (2010): 263–277.

Marcus, G.E. and Myers, F.R. "The traffic in art and culture. An introduction." In *The Traffic in Culture. Refiguring Art and Anthropology.* Edited by Marcus, G.E. and Myers, F.R. 1–51. Berkeley: University of California Press. 1995a.

Marcus, G.E. and Myers, F.R. *The Traffic in Culture. Refiguring Art and Anthropology.* Berkeley: University of California Press. 1995b.

Marres, N. "Testing power of engagement: Green living experiments, the ontological turn and the undoability of involvement." *European Journal of Social Theory* 12 (1) (2009): 117–133.

Marres, N. "As ANT is getting undone, can pragmatism help us re-do it?" In *The Routledge Companion to Actor-Network Theory.* Edited by Blok, A., Farias, I., and Roberts, C. London and New York: Routledge. 2019.

Marston, S.A. and De Leeuw, S. "Creativity and geography: Towards a politicized intervention." *Geographical Review* 103 (2013): III–XXVI.

Massey, D. "Space-time, 'science' and the relationship between physical and human geography." *Transactions of the Institute of British Geographers* 24 (1999): 261–276.

Matless, D. *The Regional Book.* Axminster: Uniform Books. 2015.

Mayhew, R.J. "Materialist hermeneutics, textuality and the history of geography: print spaces in British geography, c.1500–1900." *Journal of Historical Geography* 33 (3) (2007): 466–488.

McCormack, D.P. "Drawing out the lines of the event." *Cultural Geographies* 11 (2004): 211–220.

McCormack, D.P. "Thinking spaces of research creation." *Inflexions* 1 (2008): 1. https://www.inflexions.org/n1_Thinking-Spaces-for-Research-Creation-by-Derek-P-McCormack.pdf.

McCormack, D.P. "Thinking in transition: The affirmative refrain of experience/experiment." In *Taking-Place: Non-representational Theories and Geography.* Edited by Anderson, B. and Harrison, P. 201–220. Farnham: Ashgate. 2010.

McCormack, D.P. *Refrains for Moving Bodies.* Durham, NC: Duke University Press. 2014.

McLean, H. "Hos in the garden: Stating and resisting neoliberal creativity." *Environment and Planning D: Society and Space* 35 (2017): 38–56.

Meinig, D. "Geography as an art." *Transactions of the Institute of British Geographers* 8 (1983): 314–328.

Mercer, L. "Speculative emblematics." PhD Thesis, Royal Holloway, University of London. (In preparation).

Merriman, P., Revill, G., Cresswell, T., Lorimer, H., Matless, D., Rose, G., and Wylie, J. "Landscape, mobility, practice." *Social & Cultural Geography* 9 (2) (2008): 191–212.

Mol, A.M. *Body Multiple: Ontology in Medical Practice*. Durham, NC: Duke University Press. 2002.

Möntmann, N. *Art and Its Institutions: Current Conflicts, Critique and Collaborations*. London: Black Dog Publishing. 2006.

Mott, C. and Cockayne, D. "Citation matters: Mobilizing the politics of citation toward a practice of 'conscientious engagement'." *Gender, Place & Culture* 247 (2017): 954–973.

Mott, C. and Roberts, S.M. "Not everyone has (the) balls; urban exploration and the persistence of masculine geography." *Antipode* 46 (1) (2014): 229–245.

Mould, O. *Against Creativity*. London: Verso. 2018.

Mountz, A., Bonds, A., Mansfield, B., Lloyd, J., Hyndman, J., Walton-Roberts, M., Basu, R., Whitson, R., Hawkins, R., Hamilton, T., and Curran, W. "For slow scholarship: A feminist politics of resistance through collective action in the neoliberal university." *ACME: International E-Journal for Critical Geographies* 14 (4) (2015): 1235–1259.

Mulholland, D., Lorimer, H., and Philo, C. "Resounding; an interview with Drew Mulholland." *Scottish Geographical Journal* 125 (3–4) (2009): 379–400.

Nagar, R. "Storytelling and co-authorship in feminist alliance work: Reflections from a journey." *Gender, Place and Culture* 20 (1) (2013): 1–18.

Neate, H. and Craggs, R. (ed). *Modern Futures*. Axminster: Uniform Books. 2016.

Newman, B. "Authorising geographical knowledge: The development of peer review in *The Journal of Royal Geographical Society, 1830–1880.*" *Journal of Historical Geography* 64 (2019): 85–97.

Neidderer, K., Biggs, M., and Ferris, M. (2006) "The research exhibition: Context, interpretation and knowledge creation." In *Designs Research Society International Conference Proceedings: Wonderground*. Edited by Friedman, K., Love, T., Corte-Real, E., and Rust, C. Vol. 0120. IADE.

Nimkulrat, N. "Situating creative artefacts in art and design research." *FORMakademisk* 6 (2) (2013): 1–16.

Noble, K.A. *Changing Doctoral Degrees: An International Perspective*. Buckingham: Society for Research into Higher Education. 1994.

Norcliffe, G. and Rendace, O. "New geographies of comic book production in North America: The new artisan, distancing and the period social economy." *Economic Geography* 79 (3) (2003): 241–263;

Nussbaumer, S.U. and Zumbuhl, H.J. "The little ice age history of the glacier des Bossons (Mont Blanc massif, France): A new high-resolution glacier length curve based on historical documents." *Climatic Change* 111 (2012): 301–334.

Oberhauser, A.M. and Caretta, M.A. "Mentoring early career women geographers in the neoliberal academy: Dialogue, reflexivity, and ethics of care." *Geografiska Annaler: Series B, Human Geography* 101 (1) (2019): 56–67.

Obrist, H.-U. and Vanderlinden, B. *Laboratorium*. Colonge: Dumont. 2001.

O'Dell, K. "Displacing the haptic: Performance art, the photographic document and the 1970s." *Performance Research, A Journal of the Performing Arts* 2 (1) (2014): 73–81.

Offman, J. (ed). *The Studio: Documents of Contemporary Art*. London and Cambridge, MA: White Chapel and MIT Press. 2012.

Ogborn, M. and Withers, C. *Geographies of the Book*. Farnham: Ashgate Publishing. 2010.

Ogborn, M. and Withers, C. *Geographies of Nineteenth-Century Science*. Chicago: University of Chicago Press. 2011.

Olsson, G. *Birds in Eggs/Eggs in Birds.* London: Pion. 1980.

Osborne, T. "Against 'creativity': A philistine rant." *Economy and Society* 32 (4) (2003): 507–525.

Osborne, P. *Anywhere or Not at All: Philosophy of Contemporary Art.* London: Verso, 2013.

O'Neill, P. and Wilson, M. (eds). *Curating Research.* London: Open Editions. 2015.

Overend, D. and Lorimer, J. "Wild performatives: Experiments in rewilding at the Knepp wildland project." *GeoHumanities* 4 (2) (2018): 527–542.

Overend, D., Lorimer, J., and Schreve, D. "The bones beneath the streets: drifting through London's quaternary." *Cultural Geographies* (2019), online first.

Paglen, T. "Experimental geography: From cultural production to the production of space." In *Experimental Geography.* Edited by Thompson, N. and Independent Curators International. 22–33. New York: Melville House. 2009.

Pain, R. "Impact: Striking a blow or walking together?" *ACME: An International e-Journal for Critical Geographies* 13 (1) (2014): 19–23.

Pain, R. and Kindon, S. "Participatory geographies." *Environment and Planning A* 39 (12) (2007): 2807–2812.

Paltridge, B., Starfield, S., Ravelli, L., and Nicholson, S. "Doctoral writing in the visual and performing arts: Issues and debates." *International Journal of Art & Design Education* 30 (2) (2011): 242–255.

Paltridge, B., Starfield, S., Ravelli, L., and Nicholson, S., "Doctoral writing in the visual and performing arts: Two ends of a continuum." *Studies in Higher Education* 37 (8) (2012): 989–1003.

Paltridge, B. Starfield, S., Ravelli, L.J., and Tuckwell, K. "Change and stability: Examining the macrostructures of doctoral theses in the visual and performing arts." *Journal of English for Academic Purposes* 11 (4) (2012): 332–344.

Park, C. *Redefining the Doctorate.* 2007. https://www2.le.ac.uk/departments/doctoralcollege/about/external/publications/redefining-the-doctorate.pdf.

Parr, H. "Collaborative film-making as process, method and text in mental health research." *Cultural Geographies* 14 (2007): 114–138.

Parrott, F. and Hawkins, H. "Six voids." In *A Place More Void.* Edited by Kingsbury, P. and Secor, A. 2020. Forthcoming.

Parrott, F. and Hawkins, H. (forthcoming) Conversations in Caves, *Leonardo.*

Patchett, M. and Mann, J. "Five advantages of skill." *Cultural Geographies* 25 (1) (2018): 23–29.

Pearson, M. and Shanks, M. *Theatre/Archaeology: Disciplinary Dialogues.* London: Routledge. 2001.

Pérez, M.A. "Lines underground: Exploring and mapping Venezuela's cave environment." *Cartographica: The International Journal for Geographic Information and Geovisualization* 48 (2013): 293–308.

Peterle, G. (2019) "Comics and maps?" *A CartoGrahic Essay.* http://livingmaps.review/journal/index.php/LMR/article/view/185/362 (accessed 2/1/2020).

Phelan, P. *Unmarked: The Politics of Performance.* London: Routledge. 1993.

Phillips, P. "Fieldwork: Ashness bridge." *Ecumene* 7 (1) (2000): 110–111.

Phillips, P. "Doing art and doing cultural geography: *The* fieldwork/field walking *project.*" *Australian Geographer* 35 (2) (2004): 151–159.

Pickles, J. "Radical thought-in-action: Gunnar Olsson's critique of cartographic reason." *Geografiska Annaler. Series B, Human Geography* 89 (4) (2007): 394–397.

Pile, S. *City A-Z: Urban Fragments.* London: Routledge. 2000.

Pinder, D. *Visions of the City - Utopianism, Power and Politics in Twentieth Century Urbanism.* Edinburgh: Edinburgh University Press. 2005.

Pink, S. *Doing Visual Ethnography*. London: Sage. 2012.

Powell, R.C. and Vasudevan, A. "Geographies of experiment." *Environment and Planning A*, 39 (8) (2007): 1790–1793.

Pratt, A. "Its Space Jim, but not as we know." 2002. http://diffusion.org.uk/?p=97.

Pratt, G. and Johnson, C. *Migration in Performance; Crossing the Colonial Present*. London: Routledge. 2019.

Pred, A. *Recognizing European Modernities: A Montage of the Present*. New York: Routledge. 1995.

Rancière, J. *The Politics of Aesthetics*. London and New York: Continuum. 2004.

Raunig, G. *Factories of Knowledge, Industries of Creativity*. Boston, MA: Semiotext(e). 2013.

Raynor, R. "Dramatising austerity: Holding a story together (and why it falls apart...)." *Cultural Geographies* 24 (2017): 193–212.

Relyea, L. *Your Everyday Art World*. Cambridge, MA: MIT Press. 2013.

Rendell, J. *Art and Architecture: A Place Between*. London: IB Tauris. 2006.

Richardson-Ngwenya, P. "Performing a more-than-human material imagination during fieldwork: Muddy boots, diarising and putting vitalism on video." *Cultural Geographies* 21 (2) (2013): 293–299.

Richardson, M.J. "Theatre as safe space? Performing intergenerational narratives with men of Irish descent." *Social & Cultural Geography* 16 (6) (2015): 615–633.

Rogers, A. "Advancing the geographies of the performing arts." *Progress in Human Geography* 42 (2018): 549–568.

Rogoff, I. "Turning." November, *E-Flux*. 2008. https://www.e-flux.com/journal/00/68470/turning/ (accessed 18/12/19).

Rose, G. *Visual Methodologies*. London: Sage.

Rowe, W. "The wordless doctoral dissertation: Photography as scholarship." *Communication* (fall) (1995): 1–30.

Ruddock, J. *Navigating the Uncertainties of Art and Science Collaboration: A Series of Projects Focused on Climate Change*. 2018. https://www.academia.edu/41288196/NAVIGATING_THE_UNCERTAINTIES_OF_ART_AND_SCIENCE_COLLABORATION_A_SERIES_OF_PROJECTS_FOCUSSED_ON_CLIMATE_CHANGE (accessed 18/12/19).

Rupke, N. "A geography of enlightenment: The critical reception of Alexander von Humboldt's Mexico work." In *Geography and Enlightenment*. Edited by Livingstone, D.N. and Withers, C.W.J. 319–340. Chicago, IL: University of Chicago Press. 1999.

Rust, C. and Robertson, A. "Show or tell? Opportunities, problems and methods of the exhibition as a form of research dissemination." EAD Conference Barcelona. 2003. http://www.ub.es/5ead (accessed 18/12/19).

Ryan, J.R. *Picturing Empire: Photography and the Visualisation of the British Empire*. London: Reaktion. 2013.

Rycroft, S. "The artist placement group: An Archaeology of Impact." *Cultural Geographies* 26 (3) (2019): 289–304.

Sachs-Olsen, C. "Performing urban archives - a starting point for exploration." *Cultural Geographies* 23 (3) (2016): 511–515.

Sachs-Olsen, C. "Urban space and the politics of socially engaged art." *Progress in Human Geography* 43 (6) (2018): 985–1000.

Sachs-Olsen, C. "Performance and urban space: An ambivalent affair." *Geography Compass* 12 (12) (2018a): 1–10.

Sachs-Olsen, C "Collaborative challenges: Negotiating the complicities of socially engaged art within an era of neoliberal urbanism." *Environment and Planning D: Society and Space* 36 (2) (2018b): 273–293.

Sachs, Olsen, C. *Socially Engaged Art and the Neoliberal City*. London: Routledge. 2019.

Sachs-Olsen, C. and Hawkins, H. "Archiving an urban exploration - MR NICE GUY, cooking oil drums, sterile blister packs and uncanny bikinis." *Cultural Geographies* 23 (3) (2016): 531–543.

Scalway, H. "A patois of pattern: Pattern, memory and the cosmopolitan city." *Cultural Geographies* 13 (2) (2006): 451–457.

Schwarzenbach, J. and Hackett, P. *Transatlantic Reflections on the Practice-Based PhD in Fine Art*. New York: Routledge. 2015.

Scott, J. *Artists-in-Labs: Processes of Inquiry*. New York: Springer. 2006.

Scott, J. *Artists-in-Labs: Networking in the Margins*. New York: Springer. 2011.

Sharp, J. "Geography and gender: Feminist methodologies in collaboration and in the field." *Progress in Human Geography* 9 (3) (2005): 304–309.

Sheikh, S. "Towards the exhibition as research." In *Curating Research*. Edited by O'Neill, P. and Wilson, M. 32–46. London: Open Editions. 2015.

Sjöholm, J. *Geographies of the Artist's Studio*. London: Squid and Tabernacle. 2012.

Sjöholm, J. "The art studio as archive: tracing the geography of artistic potentiality, progress and production." *Cultural Geographies* 21 (3) (2014): 505–514.

Slager, H. "Nameless science, art and research: A journal of ideas, contexts and methods." 2 (2) (2019). http://www.artandresearch.org.uk/v2n2/slager.html (accessed 18/12/19).

Slater, H. "The art of governance." *Variant* 2 (11) (2000): 23–33.

Smith, B. *Imagining the Pacific in the Wake of the Cook Voyages*. Melbourne: Melbourne University Press. 1992.

Smith, B. *European Vision and the South Pacific 1768–1850: A Study in the History of Art and Ideas*. New Haven and London: Yale Univeraity Press. 1988.

Smith, H. and Dean, R.T. "Introduction: Practice-led research, research-led practice – towards the iterative cyclic web." In *Practice-Led Research, Research-Led Practice in the Creative Arts*. Edited by Smith, H. and Dean, R.T. 1–38. Edinburgh: Edinburgh University Press. 2009.

Springett, S. "Going deeper of flatter: Connecting deep mapping, flat ontologies and the democratizing of knowledge." *Humanities* 4 (4) (2015): 623–636.

Stafford, B. *Voyage into Substance: Art, Science, Nature and the Illustrated Travel Account, 1760–1840*. Cambridge, MA: MIT Press. 1984.

Staheli, L.A. and Lawson, V. "'A discussion of "Women in the Field': The politics of feminist fieldwork." *Professional Geographer* 46 (1994): 96–102.

Straughan, E. "A touching experiment: Tissue culture, tacit knowledge and the making of bioart." *Transactions of the Institution of British Geographers* 44 (2) (2019): 214–225.

Straughan, E. and Dixon, D. "Rhythm and mobility in the inner and outer hebrides: Archipelago as art-science research site." *Mobilities* 9 (3) (2014): 452–478.

Stephens, K. "Artists in residence in England and the experience of the year of the artist." *Cultural Trends* 11 (42) (2001): 41–76.

Svenungsson, J. "The writing artist." *Art & Research* 2 (2) (2009): 1–6.

Thrift, N. "The future of geography." *Geoforum* 33 (2002): 291–298.

Thrift, N. *Spatial Formations*. London: Sage. 2006.

Thrift, N. *Non-representational Theory: Space, Politics, Affect*. London, Routledge. 2009.

Thrift, N., Harrison, S., and Pile, S. *Patterned Ground: Entanglements of Nature and Culture*. London: Reaktion Books. 2004.

Till, K. "Artist and activist memory work: Approaching place based practice." *Memory Studies* 1 (2008): 99–113.

Tolia-Kelly, D.P. "Fear in paradise: The affective registers of the English Lake district landscape re-visited." *The Senses and Society* 2 (2007a): 329–351.

Tolia-Kelly, D.P. (2007b) "SPILL: Liquid emotion and transcultural art." *SPILL Exhibition Catalogue*. SASA Gallery.

Tolia-Kelly, D.P. "The geographies of cultural geography II: Visual culture." *Progress in Human Geography* 36 (2011): 135–142.

Tooth, S., Viles, H.A., Dickinson, A., Dixon, S.J., Falcini, A., Griffiths, H.M., ... Whalley, B. "Visualising geomorphology: Improving communication of data and concepts through engagement with the arts." *Earth Surface Processes and Landforms* 41 (12) (2016): 1793–1796.

Velios, A. "Creative archiving: A case study from the John Latham Archive." *Journal of the Society of Archivists* 32 (2011): 255–271.

Vergo, P. "The reticent object." In *The New Museology*. Edited by P. Vergo, P. 41–59. London: Reaction Books. 1989.

Vilches, F. "The art of archaeology: Mark Dion and his dig projects." *Journal of Social Archaeology* 7 (2) (2007): 199–223.

Voorhies, J. *Beyond Objecthood: The Exhibition as a Critical Form since 1968*. Cambridge, MA: MIT Press. 2017.

Wainwright, E., Barker, J., Ansell, N., Buckingham, S., Hemming, P., and Smith, F. "Geographers out of place: Institutions, (inter)disciplinary and identity." *Area* 46 (2014): 410–417.

Waite, L.J. and Conn, C. "Creating a space for young women's voices: Using participatory 'video drama' in Uganda." *Gender, Place and Culture* 18 (1) (2011): 115–135.

Walker, J. "Artist placement group (APG): The individual and the organisation. A decade of conceptual engineering." *Studio International* 191 (1976): 164.

Walker, J. and Latham, J. *The Incidental Person: His Art and Ideas*. London: Middlesex University Press. 1994.

Ward, M. "The art of writing place." *Geography Compass* 8 (2014): 755–766.

Warf, B. and Arias, S. *The Spatial Turn: Interdisciplinary Perspectives*. London: Routledge. 2009.

Warren, A. and Gibson, C. *Surfing Places, Surfboard Makers*. Honolulu: University of Hawai'i Press. 2014.

Weizman, E. *Forensic Architecture: Violence at the Threshold of Detectability*. Cambridge, MA: Zone Books, MIT Press. 2017.

Whatmore, S. "Mapping knowledge controversies: Environmental science, democracy and the redistribution of expertise." *Progress in Human Geography* 33 (5) (2009): 587–598.

Whatmore, S. and Landström, C. "Flood apprentices: An exercise in making things public." *Economy and Society* 40 (4) (2011): 582–610.

Wilk, E. "The artist-in-consultance: Welcome to the new management." *E-flux* 74 (June 2016), unpaginated.

Wilson, J. "The white cube in the black box: Assessing artistic research quality in multidisciplinary academic panels." *Assessment and Evaluation in Higher Education* 41 (8) (2015): 1223–1236.

Wisker, G., Robinson, G., Trafford, V., Warnes, M., and Creighton, E. "From supervisory dialogues to successful PhDs: Strategies supporting and enabling the learning conversations of staff and students at postgraduate level." *Teaching in Higher Education* 8 (3) (2003): 383–397.

Withers, C.W.J. and Livingstone, D.N. "Thinking geographically about 19th century science." In *Geographies of Nineteenth-Century Science*. Edited by Livingstone, D.N. and Withers, C.W.J. 1–20. Chicago, IL: University of Chicago Press.

Woodward K., Jones III, J.P., Vigdor, L., Marston, S.A., Hawkins, H., and Dixon, D.P. "One sinister hurricane: Simondon and collaborative visualization." *Annals of the Association of American Geographers* 105 (3) (2015): 496–511.

Worpole, K. "Nowhere to run, nowhere to hide." 2015. https://thenewenglishlandscape.
   wordpress.com/2015/10/07/nowhere-to-run-nowhere-to-hide/ (accessed 18/12/19).
Wylie, J. and Webster, C. "Eye-opener: Drawing landscape near and far." *Transactions of
   the Institute of British Geographers* 44 (1) (2019): 32–47.
Yusoff, K. "Landscape: Return dispersal circulation." 2006. http://diffusion.org.
   uk/?p=113 (accessed 18/12/19).

INDEX

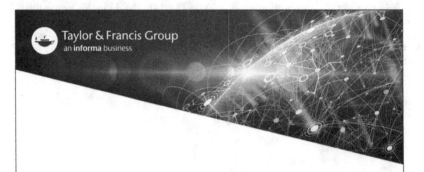

Printed in the United States
By Bookmasters